Inventions Necessity Is Not The Mother Of
Patents Ridiculous and Sublime

Inventions Necessity Is Not The Mother Of

Patents Ridiculous and Sublime

Stacy V. Jones

Quadrangle/The New York Times Book Co.

 For information, address: Quadrangle/The New York Times Book Co., 10 East 53rd Street, New York, New York 10022. Manufactured in the United States of America. Published simultaneously in Canada by Fitzhenry & Whiteside, Ltd., Toronto.

Library of Congress Catalog Card Number: 72-90461

International Standard Book Number: 0-8129-0315-3

Jacket design by: Ampersand Design, Inc.
Interior design by: Planned Production

Contents

Preface

In preparing this book, whose compilation has been far from onerous, I have had the unconscious collaboration of hundreds of fellow writers. Once their creative compositions were on paper, they were illustrated, and—after a good deal of fuss and many fees—were published as patents. These compositions are now on the shelves of an enormous library centered in Crystal City, Arlington County, Virginia, known as the Patent Office.

In thanking the authors of the patents, more commonly known as inventors, I must also express my appreciation to their professional associates, practitioners of the patent law. The advice of these attorneys often has social as well as practical value. For example, a Washington lawyer once received a letter from an inventor who asked his assistance in patenting a do-it-yourself suicide kit. The attorney responded that he did not think the government could provide protection for that invention.

Seriously, I am indebted to a number of attorneys who over the years have given me patent numbers as clues to the quaint, the diverting, the absurd and occasionally the ludicrous. One I must mention is Gustave Miller of Miami, who lent me collections of his patents and provided other help. I must also thank many Patent Office people, including Isaac Fleischmann, Oscar G. Mastin, Robert J. Rish and the recently retired Ernest A. Norwig. Closer to home I have had the valuable assistance of my wife, Margaret Crahan Jones.

The New York Times has given space to many of the items in this book as they were disclosed from week to week along with more prosaic inventions of value to industry and the consumer. The book title is derived from headlines used in *The New York Times Magazine.* Acknowledgment is also due *Science Digest* for permission to incorporate or paraphrase certain material.

No claim is made that this book offers the one and only possible collection of oddball inventions. There are literally millions of patents, and a fresh thousand or so can be expected every Tuesday. Also, another collector might well be intrigued by other things.

A book such as this, whose purpose is entertainment rather than the advancement of science, is not likely to provide source material for serious research. Many of the inventions mentioned exist only as concepts and have never taken physical form. But readers who want to get complete copies of the patents can do so with a little trouble and expense. As an aid, an effort has been made to show the year of issuance for each patent mentioned. For any whose number is not given in the margin, the suggested procedure is: Go to a large public or university library and ask to see the annual *Index of Patents* for the year of issue of the patent. If that volume is available, the patent number will be shown under the inventor's name. Send the number and a check or money order to the Commissioner of Patents, Washington, D.C. 20231. The copies are 50 cents each, except for design patents which are 20 cents apiece. For very old patents, a visit to the Patent Office or payment of a search fee may be necessary.

It has been pleasant talking with you about my hobby.

Stacy V. Jones

Falls Church, Virginia
March, 1973

Inventions Necessity Is Not The Mother Of
Patents Ridiculous and Sublime

1

What's Funny about That?
Invention Is a Serious Business

The inventive spirit may generate anything from nuclear fusion, achieved after years of scientific research, to a new cocktail hit upon by picking up the wrong bottle. Many discoveries are never used and go unnoticed or are carefully hidden as corporate trade secrets.

We are lucky that society has recorded millions of these ideas in patent offices. It is not surprising that in the vast sea of patents the incongruous often comes to the surface. Inventions outside their usual frame of reference seem absurd. Examples are a computer for guessing the outcome of a horserace, a camera apparatus for fitting clothes and a sleeping hammock for mink. Sometimes we feel superior because times have changed—the outdoor toilet heated by a kerosene stove no longer seems essential to the household.

Patents are requested not only by those with a selfish desire for profit gained by enforcement of the limited monopoly they contain but also by those with other human motives, some of them generous. Love of children may have motivated the invention of knives and forks equipped with whistles, and man's sympathy for animals is illustrated by a machine that a dog can scratch himself with.

Inventors are human. In this country patents are never granted to computers or corporations but to people, without regard to age, sex or race. Youngsters are among the patentees and in 1972 an award was made to a scientist just after his 100th birthday. So it is not surprising that patented inventions exhibit human characteristics, particularly humor, sometimes intentional, but often accidental.

An early example may be drawn from England, from whose patent system the American one was derived. In 1718 George I graciously granted to one James Puckle, a gentleman of London, a patent for a machine gun. The design showed a single barrel and revolving chambers that were brought one at a time into line with the breech for firing. The significant point is that Mr. Puckle provided square bullets for use against the Turks and round ones for Christians.

The first United States patent, signed by George Washington in 1790, went to Samuel Hopkins of Philadelphia for an improvement in "the making of Pot ash and Pearl ash by a new Apparatus and Process." By 1973, 3,700,000 patents had been issued and new ones were coming in at the rate of well over 1,000 a week, bearing such solemn titles as Azimuth Interpolator and Phase Detector Using Logic Gating Circuits.

The list is 99-44/100 percent prosaic but about one in a thousand strikes a lighter note. The one-in-a-thousand makes it worthwhile to peruse the Patent Office *Official Gazette*, a thick weekly magazine containing an abstract and usually a drawing for each patent, with the inventor's name and his address, or at least his hometown. An ambitious researcher can study copies of the earlier patents in the Patent Office files at Crystal City.

A patent copy is a stapled collection of printed sheets about letter size, usually containing reproductions of drawings. The original, formal grant or master copy, which goes to the inventor or assignee, has attached to it an impressive certificate bearing the facsimile signature of the Commissioner of Patents.

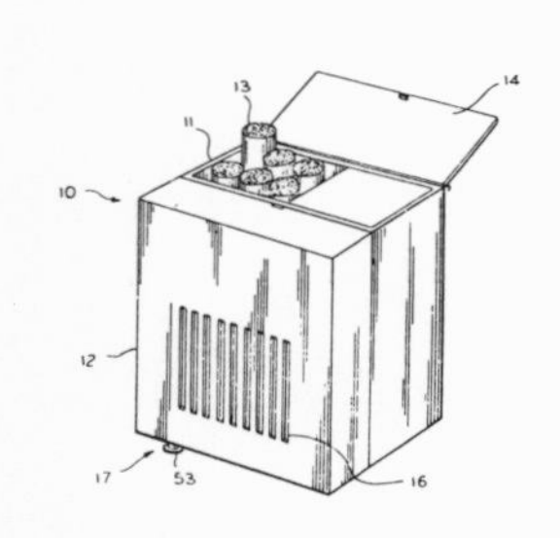

Fig. 1.1:
The cigarette package that coughs.
(Patent 3,655,325)

The patent may have meaning in some cases only to technicians skilled in a particular art, but in other cases the casual reader may easily grasp the color or sound of the concept. For example there is a pseudo-cigarette package that starts coughing loudly when someone picks it up. (Figure 1.1) Lewis R. Toppel of

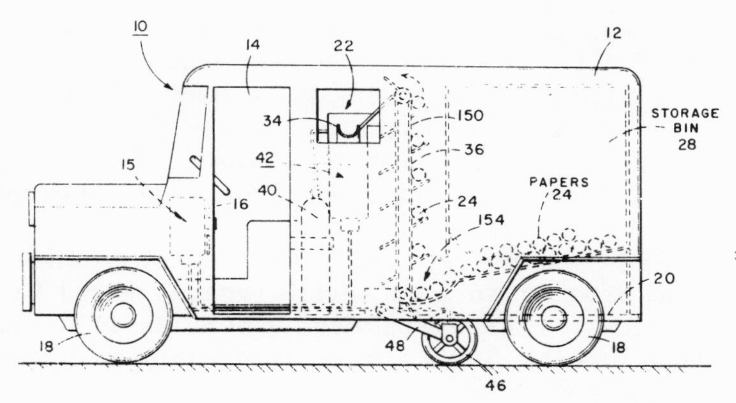

Fig. 1.2:
Truck throws newspapers into the yard.
(Patent 3,653,538)

Madison, Wisconsin, got this patent in 1972. The so-called "Smoking Deterrent" was intended to help cure the habit and also to sell as an advertising novelty. An intriguing picture is offered by another 1972 patent issued to Robert L. Lamar of Houston, Texas for a truck to be driven down a residential street throwing newspapers right and left into subscribers' yards, under specific orders from a punched tape. (Figure 1.2)

A useful machine for the food trade may entertain those in other professions. In 1953 Lindsay H. Browne of Westport, Connecticut patented a jelly squirter for filling doughnuts. Four years later, Edwin I. Groff of West Reading, Pennsylvania got a patent for a pretzel-twisting machine. And even the automobile trade may have been surprised by a gadget for motorists disclosed in 1964 by Rafael D. Bonnelly of New York—a combined car bumper and bottle opener.

In that year a professional entertainer offered fun for children. Paul Winchell, the ventriloquist, patented a mask in the shape of a human dummy upside down. As a supplement he suggested a mirror in which the child could see the figure rightside up. Like other patents which cover the mechanical, chemical or electrical structure of an invention, it gave him the right to exclude others from making, using or selling the mask throughout the United States for 17 years. (Figure 1.3)

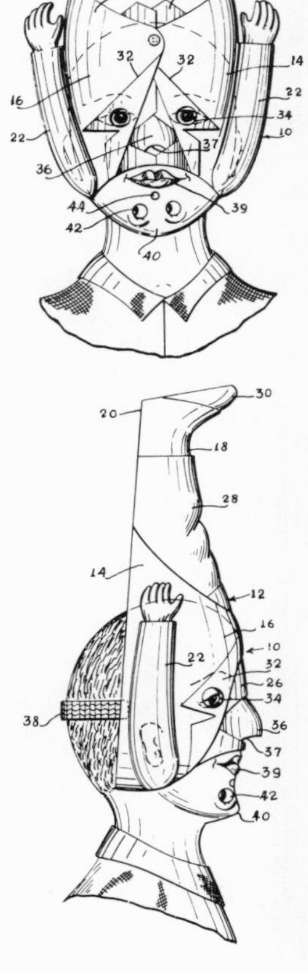
Fig. 1.3:
Ventriloquist designs mask for kids.
(Patent 3,129,001)

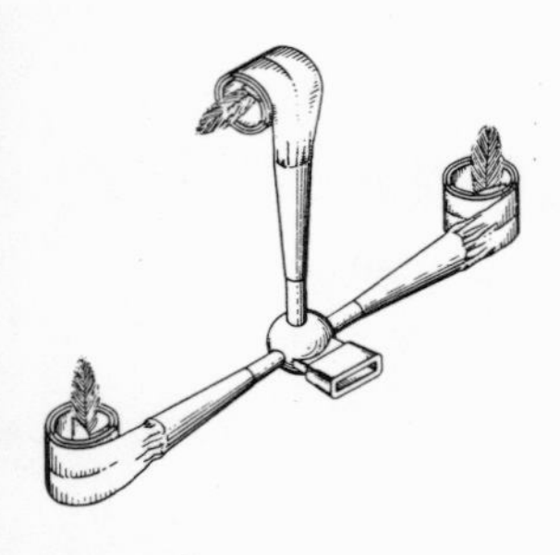

Fig. 1.4:
Danny Kaye offers entertainment.
(Patent D.166,807)

Fig. 1.5:
Divining rod explores for oil.
(Patent D.192,048)

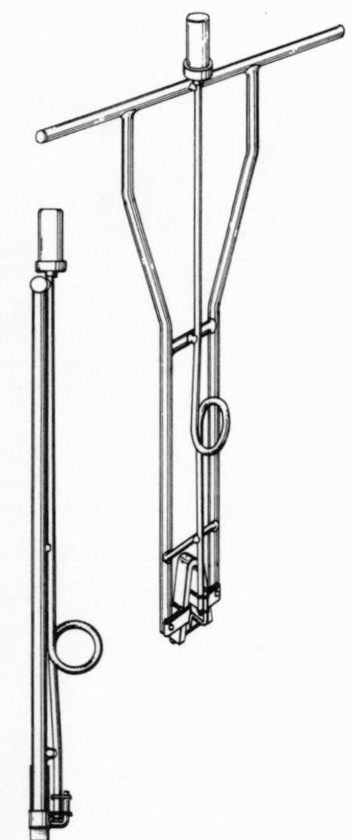

Design patents are a limited class that relate only to the appearance of the subject and are issued for 3½, 7 or 14 years. As early as 1952 Danny Kaye and a co-inventor, Edward Dukoff, got a design patent for a blowout toy. It provided three "snakes" that unrolled when inflated, with feathers at the end for tickling noses. There was also a device to deliver a mild Bronx cheer. (Figure 1.4)

As he was not required to prove to the examiners that it would work but merely what it looked like, Roy Rickard, a house painter and decorator in Toledo, Ohio was able to get a design patent in 1962 entitled Divining Rod. He told a reporter it was a frame of copper tubing with a cross bar at the top to be held in the hands, and at the bottom it had a tip that fluctuated when there were hydrocarbons below the ground. The rod was intended as a prospecting tool for oil and gas. (Figure 1.5)

The Patent Office never knowingly grants patents for perpetual motion, but a number have slipped by under other titles. John Sutliff, Sr., of Huntsville, Missouri received one in 1882 for what was simply titled "Motor." It had a water tank, a lever, a bulb filled with air and several moving parts but no fuel or other source of power. (Figure 1.6)

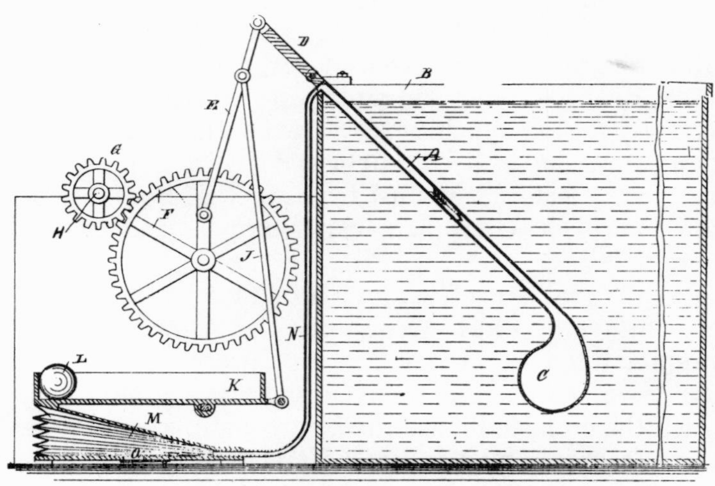

Fig. 1.6:
Motor to provide perpetual motion.
(Patent 257,103)

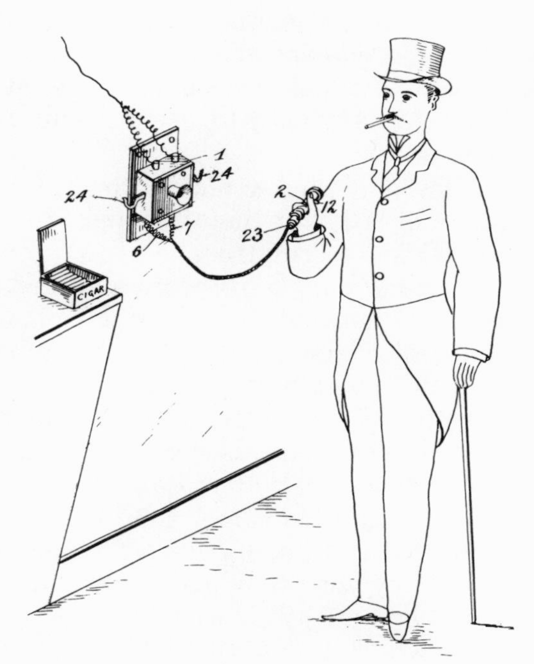

Fig. 1.7:
Lighting a cigar properly in 1904.
(Patent 774,624)

Many patents are amusing merely because of their age. Examples are an ear trumpet carried in a tall hat, a method of winding a clock with rainwater, a coyote alarm with a scarecrow that shoots blanks, a flask for prohibition agents to use in gathering evidence and mustache guards to facilitate eating. The guards may have to be revived, of course, if men's facial styles get any hairier. The electric cigar lighter shaped like a wall telephone was doubtless an appealing novelty in 1904, but the formal drawings in a patent granted to James Waters and Frederick V. Thompson of Philadelphia are quaint today, as Figure 1.7 shows.

Inventors can be witty. Thomas Windell of New Albany, Indiana displayed humor in his application for a patent issued in 1860 on glass tombstones in which inscriptions could be pressed. In a drawing of one monument (Figure 1.8) he inscribed his own name, with a verse ending:

A curious fact
It has sometimes been said
That he made it while living
But enjoys it while dead.

Fig. 1.8:
Humor on the inventor's tombstone.
(Patent 28,029)

It should be mentioned that like all patents issued 17 or more years ago Mr. Windell's has expired and his tombstone invention has gone into the public domain for anybody to use without financial obligation.

Now and then a notable patent concerns government or a government leader. In 1970 Mrs. Alba W. Hicks, a sculptress, and her son John—both of Bethesda, Maryland—were awarded a design patent for a statuette of Richard M. Nixon, one of a series of Political Pixies sold in Washington and elsewhere. The President is shown in a thoughtful pose with a finger on his upper lip. As manufactured, the statuette carried this slogan on its base, "All quiet on the Potomac." (Figure 1.9)

Fig. 1.9:
Patented Nixon statuette.
(Patent D.216,807)

Americans who feel that it takes too long for a letter to reach them from across the country or across town may wish they had the mail service system patented in 1901 by George Alfred Owen of Springfield, Massachusetts. In 13 sheets of drawings and some 10,000 words of text he describes a mechanized service that accepts letters in unmanned curbside stations and transfers them to overhead conveyors, in which they are automatically transported to the central office. The same conveyors deliver sorted mail to letter-boxes in the curbside stations. The pressure of a piece of mail on a switch at the bottom of a box sends a signal to the addressee's home, inviting him or her to call for it. When the addressee doesn't want to be bothered by signals, the wires can be disconnected. (Figure 1.10)

Fig. 1.10:
Automatic mail service offered in 1901.
(Patent 677,423)

Now and then a modern inventor, with the collaboration of his attorney, gives the fun treatment to an ordinarily prosaic subject—such as ways to poison insect pests. Donald M. Stout, research director for a St. Louis laboratory, invented psychedelic decoys. His application for a patent granted in 1972 said in part:

> The invention relates generally to insect control and particularly to a method and means of luring houseflies (*Musca domestica*) into the jaws of death.
>
> Many a fly poison has been proposed—and most are quite effective, provided the fly is willing to cooperate by eating or drinking that of which humankind wants them to partake. On the other hand, flies are smart. At least they often appear to deliberately shun that which humankind has prepared especially for them, while flocking to that which humankind has prepared for itself.
>
> The object of the invention is therefore to draw flies—and having drawn them, to speed their demise. . . .
>
> In the lure of my invention, the fly appeal of both color and reflected ultraviolet light is combined with graphic silhouettes resembling flies. Preferably the graphic silhouettes are pictorial illustrations of flies in postures which, when assumed by live flies, seem to make other flies "nosey" and to arouse an instinctive response in them. However, substantially triangular black, or at least dark, spots of substantially fly-size are about as effective. The two postures which appear to have the greatest portent for enticing other flies to join the party are the posture of copulation (which lures non-participating, but jealous, male flies) and the posture of feeding (which lures the ever-hungry species regardless of sex). Accordingly, my invention contemplates the provision of a film or integument of paper, cloth or other material preferably having at least one hole extending through it, coated on the obverse side with a material which is luminescent or a good reflector of ultraviolet light, and on the reverse side with fly toxin-containing food, and also provided on the obverse side, adjacent the hole, with life-size graphic silhouettes of the type aforesaid. Such a film, when suspended so that both sides are fly-accessible, allures flies to its obverse side where they join their illusory kind in pursuit of whatever they are looking for to satiate their whetted appetites,

and failing to find real satisfaction on the obverse side, crawl through the hole or over the edge to investigate the reverse side where they find abundant food of which they can partake freely and without restraint by the foreknowledge that their pleasure is their poison.

Edward A. Katz of Brooklyn, who in 1972 patented a sober device to enable a person with an artificial hand to hold a violin bow, amused himself with a concept of cork cuffs for pants. The cuffs were to float the trouser ends above any water in the wearer's path. In a mock description he said, "The cork cuff will be of great advantage to the user, especially in poorly paved areas and may also be of benefit to the sportsman and hunter who must sometimes run through swamps or marshy places. It might be constructed of cork, styrofoam or any other material of a buoyant nature." Although a patent promoter told him it was a feasible idea with commercial potential, Mr. Katz stopped short of a patent application.

Among the ardent collectors of offbeat patents—including the bathroom and bedroom kinds—are Patent Office officials and patent attorneys. Many a lawyer keeps a drawerful for the entertainment of clients and friends. But both the examiners and the lawyers are quick to resent any implication that inventors are screwballs.

This observer, who has met hundreds of inventors and talked to other thousands by telephone, has often been amazed and fascinated by their conceptions but he could not point to one screwball among them even if he dared. The American inventor deserves great credit for the country's advancement in the industrial arts and personifies its imagination, ingenuity and sense of humor.

2

The Little Dears
Whack Them—But Gently

Patentees have always exhibited a broadly benevolent attitude toward children, looking for ways to help them eat, sleep, keep well, learn and play. There is even a tendency to spoil them a little by sparing the rod. As an example of kindly discipline, take the paddle patented in 1953 by George F. Jorgenson of Norfolk, Virginia which has a handle that breaks if papa spanks too hard. The blade of the paddle is round and large enough "to contact a substantial area of the rump of a child."

The safety feature is provided in the handle, which is jointed and held together by strips of tape or metal so as to give way if the swat is too powerful. (Figure 2.1) But the head can't fly off and injure anyone; it dangles.

In 1971 Dewey J. Gordon of Los Gatos, California patented what he called an educational device—a paddle to be hung on the wall, carrying a removable sign. The sign displayed in the patent is "Mama's Little Helper," designed to convey a message of approval. The sign can be pulled off if the parent wants to in-

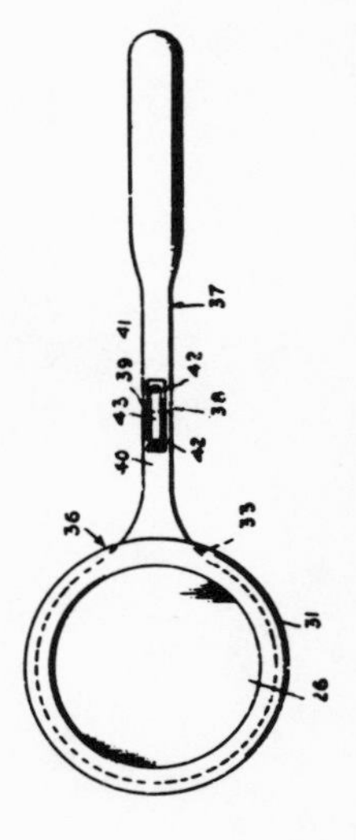

Fig. 2.1:
Spanker breaks if you swat too hard.
(Patent 2,645,488)

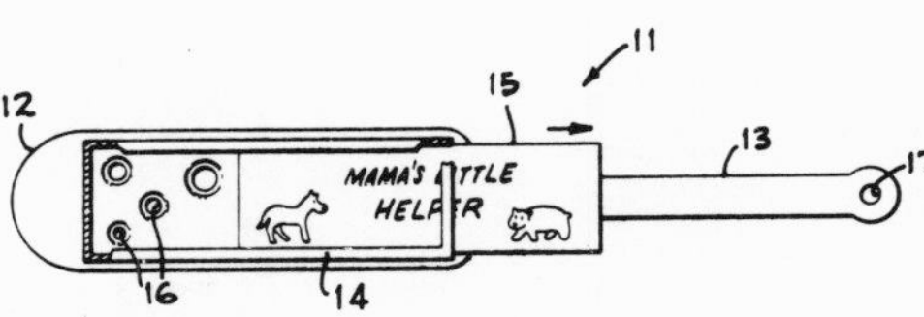

Fig. 2.2:
You better help Mama or else—.
(Patent 3,597,861)

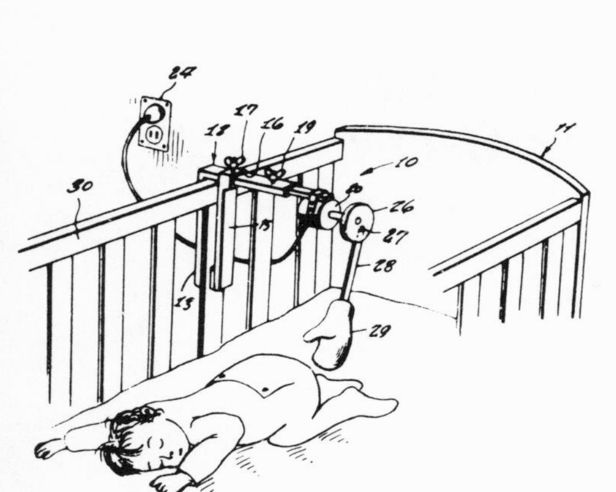

Fig. 2.3:
Patting the baby to sleep.
(Patent 3,552,388)

Fig. 2.4:
Burp seat helps with that gas.
(Patent 3,071,410)

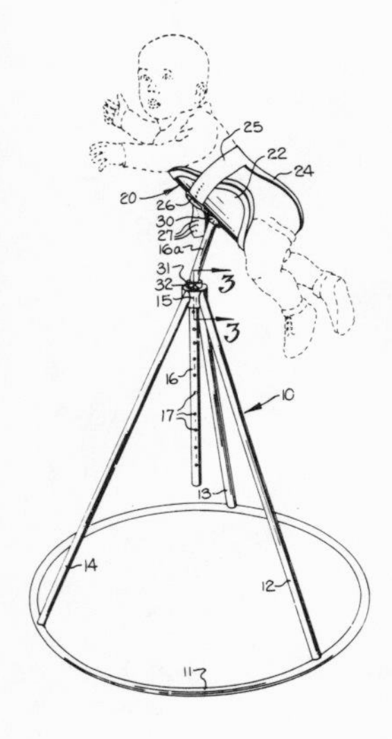

dicate disapproval; removal uncovers holes in the paddle, threatening discipline and pain. But if the paddle is actually used for spanking, a pad on the underside prevents injury to the child. (Figure 2.2)

Advancing technology offers other aids to parents. A baby-patting machine patented in 1971 by Thomas V. Zelenka of Lemoore, California takes over an onerous bedtime job. It is described as "a device for patting a baby to sleep by means of periodic pats upon the rump or hind part of the baby." With the infant lying on his or her stomach in a crib, an electrically operated mechanical arm swings a well-padded glove regularly against the designated parts of the anatomy. (Figure 2.3)

Another patting aid is entitled simply, "Baby Burp Seat." It supports an infant in a comfortable position that allows freedom of movement, gives a full view of the surroundings and aids in the release of stomach gases. The seat or saddle is elevated on a framework at about the height of an old-fashioned dictionary stand. The child rests on stomach and chest, at approximately the angle he would occupy on mother's shoulder during the conventional patting operation. As can be gleaned from a 1963 patent awarded to Glenn D. Gaskins of Clinton, South Carolina mother can go about her housework, pausing occasionally to swing the seat around, thus changing the baby's outlook and keeping him mentally occupied, as well as free from the threat of colic. (Figure 2.4)

Various means have been devised to encourage children to eat. One was "Hungry Piggy," patented in 1948 by Eoina Nudelman of Chicago. When the youngster lagged, mother urged him to give Piggy a big bite and then take one himself. At a touch of the spoon Piggy's mouth opened and the animal appeared to swallow. But unnoticed by the youngster, the cereal that Piggy ate slipped back into the bowl. By this forgivable deception, many a parent got a reluctant child to consume all the Krackly Flakes and do it promptly. (Figure 2.5)

Piggy joined many an American family and would probably have lived happily ever after had it not been for the advent of a rival, "Puppy-Tu." Puppy also sat on the edge of a cereal dish and swallowed nutriment that found its way back into the plate. The Piggy people (Topic Toys) sued the Puppy people (Crest Specialty) for infringement and won in the lower courts. But the United States Supreme Court reversed the decision, in effect recognizing Puppy as a legitimate competitor.

James Greco of Chicago recognized the appeal that music has for children. In 1955 he patented forks and spoons with handles containing whistles. Each implement is actually a fife, and by moving his fingers over holes drilled in the top of the handle, the holder can play simple tunes. Perhaps anticipating trouble with a child who tries to blow while chewing, Mr. Greco suggests that sometimes mother should do the playing.

Fig. 2.5: Hungry piggy helps children eat. (Patent 2,455,266)

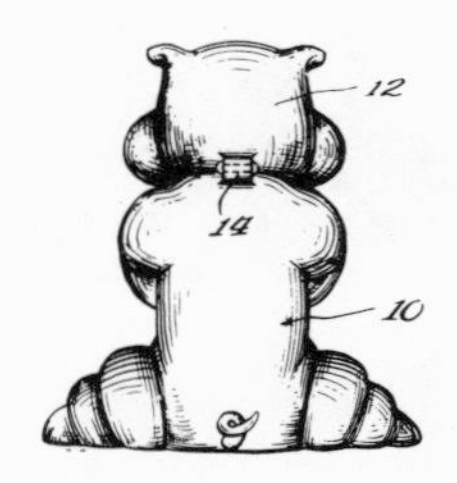

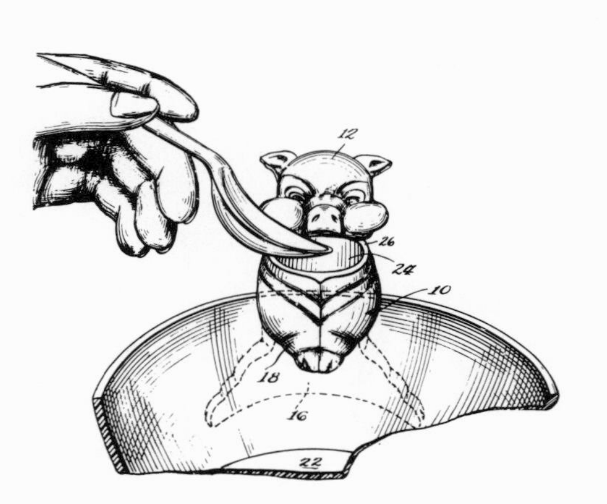

In 1905 Louis Perotti of Newark, New Jersey patented an automatic cradle that starts rocking and playing music when an infant cries and stops when the weeping stops. Through his two patents, Mr. Perotti provides in one fixture not only a clockwork motor and a music box but a clothing storage drawer, a weighing scale and a bathtub.

To induce sleep there is also the musical mattress for which Jose E. Muzaurieta of Santurce, Puerto Rico got a patent in 1968. Mother winds a music box and turns it on at bedtime. Nearby she may place one of the night lamps patented in 1957 by Ray L. Decker of Denver, which makes an endless line of sheep jump over a fence.

Then there is the musical potty. The training closet patented in 1957 by Frank Headlee of Searcy, Arkansas switches on a music box when weight is placed on the seat. The general object is to provide an appliance that is pleasing and attractive to children and is adapted to play nursery rhymes or other baby melodies. The next year Harriet Y. Clough of Meadville, Pennsylvania got a design patent for a diaper equipped with a pair of pistol holsters. (Figure 2.6)

As a permanent record of the early years Dr. Clyde J. Brauer, a dentist from Richmond, California, devised a small pair of upper and lower plates for the collection and display of a child's milk teeth. Father and mother are to gather them as they come out and stick them in place with wax. Years later, when son comes back from college—complete with permanent teeth and long hair—he can see his milk set grinning at him from the mantel, just to the right of the metalized pair of his baby shoes.

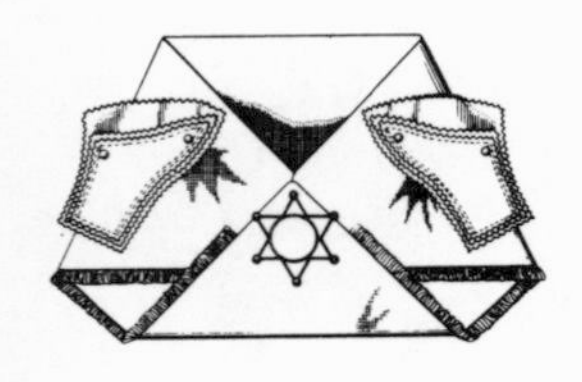

Fig. 2.6:
Diaper equipped with holsters.
(Patent D.181,838)

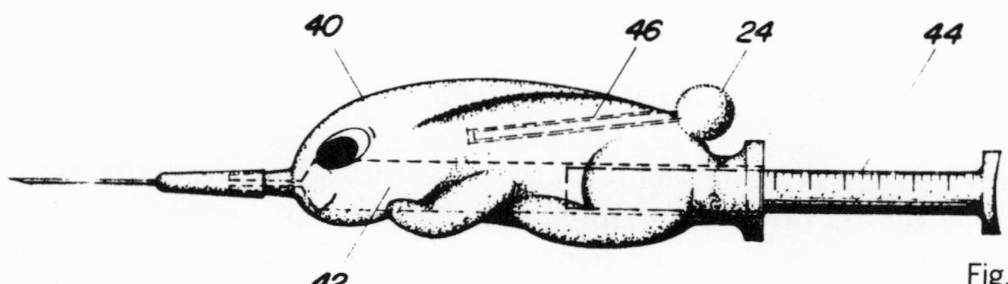

Fig. 2.7:
Bunny makes injections nothing but fun.
(Patent 3,299,891)

Another California dentist, Dr. Robert L. Smeton of Twenty-Nine Palms, has tried to make treatment less scary for children. His 1967 patent pictures a hypodermic syringe that looks like a rabbit. When pulled out, the rabbit's tail proves really to be a stick with a cotton swab. The dentist deadens pain in the child's gum by applying local anesthetic with the bunny tail. After he injects the medication, the dentist removes the needle, inserts a new swab, cleans up the rabbit and presents it to the young patient as a reward for being good. (Figure 2.7)

Fig. 2.8:
Friendly bear in the doctor's office.
(Patent D.221,938)

A large and benevolent-looking teddy bear dominates an examination table in a Long Island chiropractor's office. Dr. Anthony D. Valente of Copiague has found that the stuffed animal for which he was granted a design patent removes children's fears. The bear can sit up or lie face down. If teeth are to be checked, the youngster uses the upright bear as a backrest. And if the spine is to be examined, the bear is prostrate with the child on top. (Figure 2.8)

A considerate barber with a shop in Hampton, Virginia invented a puppet cloth for young customers to wear while their hair is being cut. A boy can insert his hands from beneath the cloth into a pair of puppets and play with them while he is being trimmed. For his own shop, Wayne Sesco, Jr. designed cloths equipped with Batman and Robin. His patent also covers a bib form for home use. (Figure 2.9)

Fig. 2.9:
Puppet play in the barber's chair.
(Patent 3,308,479)

There are certain mechanical aids for learning. Richard C. Primeau of Vancouver, British Columbia offers the student practical means of keeping one eye on the blackboard or the teacher and the other on his notebook. One side of his bisight spectacles is fitted with compensating lenses and the other with mirrors. The wearer can see with his left eye the papers on his desk but his right eye takes in distant objects.

In the same year, 1962, James F. Seligmann of New York City patented a rhyming device that helps a poet complete a word to match the end of an earlier line—such as grand and land. Many years before, in 1940, Jose N. Gonzales of Salamanca, Mexico protected by an American patent an apparatus to mechanize the conjugation of English verbs. Slotted cards and printed forms are provided instead of mental processes for decisions about such things as the preterit, the passive, auxiliaries, subjunctives and present participles.

In the livelier field of the dance, several instructional aids are offered the student. An apparatus worn like a wristwatch, patented in 1954 by John E. La Marr of Torrance, California teaches dancing by displaying the proper steps on little revolving charts. Perhaps more appealing is the mechanical figure of a young woman, with one arm to be hung around the shoulder of a living male partner. A patent issued in 1921 to Sidney E. Feist of Brooklyn explains that the man's hand fits in a strap and buckle at the back of the figure's waist. The lady has one leg extending rearward and ending in a wooden ball that rolls across the floor. The male pupil learns to lead and to avoid other couples in the ball-room, gradually mastering the maneuvers that a good dancer must acquire. (Figure 2.10)

Fig. 2.10:
The dummy that's nice to dance with.
(Patent 1,378,669)

Toys, of course, are a major contribution the patent system has made to youth, and a great deal of thought has gone into dolls. There are dolls that grow, walk, swing, sleep, smile, wet, cry, say "mama," squirm, blink, dance, wave, flop, slap, hug and kiss. Perhaps a landmark was reached in 1952 with the burping doll patented by Muriel V. Graham of Kent, Washington. When the rubber body is squeezed low-pressure air is carried up from the stomach through a rubber tube and discharged in the upper throat with the sound of a natural belch. (Figure 2.11)

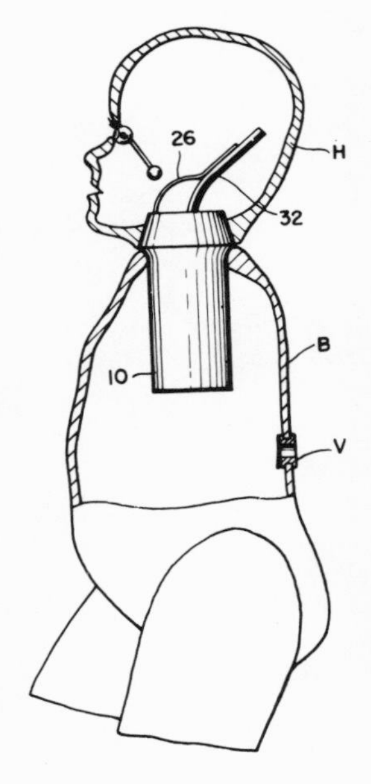

Fig. 2.11:
And the doll with the natural burp.
(Patent 2,606,399)

For kids who like to play surgery, Horace J. Munson of Fairfield, Connecticut patented in the same year a doll with removable colored plastic stomach, lungs, heart, liver, kidneys and intestines. He commented that his invention had educational value, as the young doctor learned where the organs belonged by reassembling them on pegs.

Some inventors of toys are distinguished in other fields. A flying doll was patented in 1925 by Orville Wright, who about 20 years before had been, with his brother Wilbur, inventor of the airplane. The doll is caused by a spring to fly through the air, somersault and land on a swinging frame, to which it clings with wire arms. Mr. Wright assigned the patent to the Miami Wood Specialty Company of Dayton, Ohio.

John R. Dos Passos, novelist, essayist and poet, and three friends invented a pistol that blows soap bubbles. Their bubble gun, patented in 1959, has a muzzle ending in a ring. When the ring is dipped into soapy water, a film forms across it and a blast of air sends out a stream of bubbles like bullets.

Youth is no bar to getting a patent. A number of children have received the grants, although through some coincidence their fathers are usually patent attorneys. Probably the youngest was Robert W. (Buddy) Patch of Chevy Chase, Maryland who at age 5 signed his application with a large cross and had his mark witnessed by a notary public. The patent, which he received in 1963 (when he had turned 6), was for a convertible toy truck. He had made his working model of old shoe boxes, bottle caps, Scotch tape and nails.

3

Invention with Sex Appeal
And How to Grind in a Dimple

The patent records are marked by much evidence of consideration and courtesy toward the gentler sex. Some of the gifts to womanhood made during the nineteenth century may be particularly striking in this modern era of miniskirts and pantyhose.

In the 1880s, carriages were high. When a lady lifted her voluminous skirt and climbed into one after leaving the general store, the boys in front of the livery stable across the street were afforded considerably more than a glimpse of her petticoat—they got a flash of ankle.

William Marquis Moore of Empire City, County of Clear Creek, Colorado received a patent in 1886 on his carriage-screen "for screening from the view of the bystanders the limbs of persons entering the carriage or descending therefrom." On each side of the buggy and below the body was a roller about which was wrapped a screen long enough to reach nearly to the

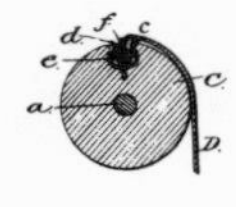

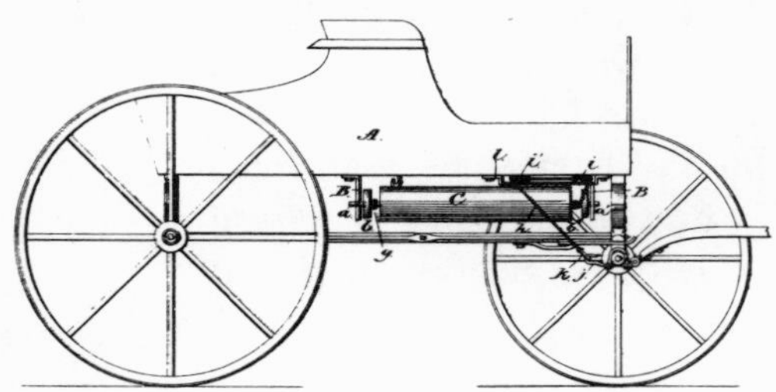

Fig. 3.1:
Carriage with modesty screen.
(Patent 346,857)

ground. The lower edge was hemmed, fitted with a bar of wood and decorated with a fringe. When the front axle was swung around, not only was the intervening wheel moved out of the lady's way but the curtain was dropped. (Figure 3.1)

Mr. Moore permitted himself a commercial thought. He suggested that the screens might carry an advertising message. Thus the livery stable crowd would be treated not to a display of hosiery, but to words extolling the merits of a brand of whiskey or tobacco.

Of equal appeal in the late nineteenth century was the self-tipping hat. What gentleman has not been embarrassed on occasions when, with his arms full of bundles, he has been unable to salute a lady? James C. Boyle of Spokane, Washington met the problem head-on.

The patent awarded him in 1896 for his saluting device said:

> This invention relates to a novel device for automatically effecting polite salutations by the elevation and rotation of the hat on the head of the saluting party when said person bows to the person or persons saluted, the actuation of the hat being produced by mechanism therein and without the use of the hands in any manner. . . .
>
> To carry into effect the broad feature of this invention, which comprehends the automatic elevation and

Fig. 3.2:
Self-tipping hat for man with his arms full.
(Patent 556,248)

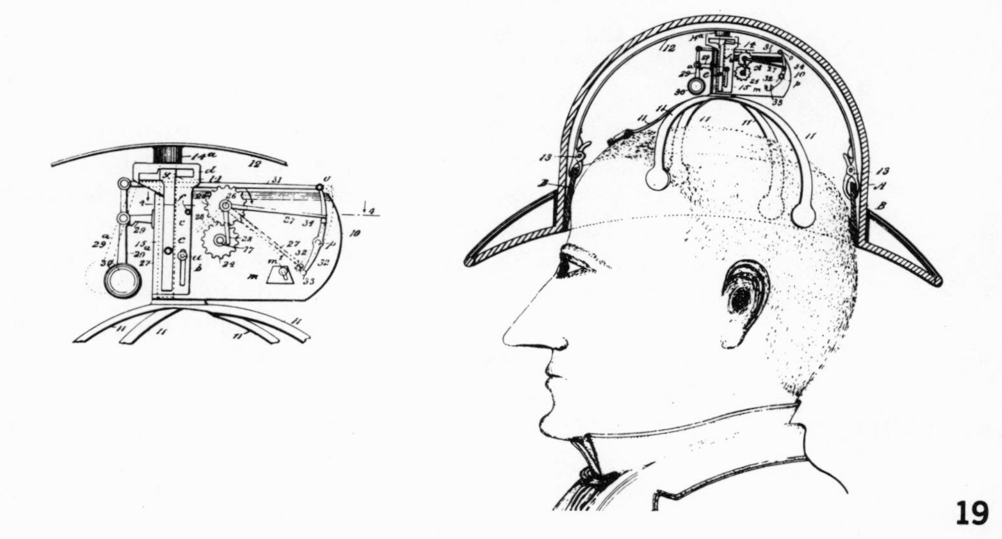

> rotation of a man's hat to effect a unique salutation, I preferably employ mechanism held in a case removably clamped on the head of the wearer of the hat, while the hat is detachably secured to the working parts of the device that raise the hat, completely rotate it and deposit it correctly on the head of the wearer every time said person bows his head and then assumes an erect posture, all parts of the novel device being completely inclosed in and concealed by the hat.

To hold the derby and mechanism in place, a number of curved springfingers gently clasped the head of the wearer. He wound the clockwork before he put the arrangement on his head. Then when he bowed to someone, a weightblock shifted and started the machinery. The hat tipped forward, then swung in a circle and settled back in place to the amazement of onlookers.

Besides courtesy, Mr. Boyle (like Mr. Moore) thought of the advertising angle. He said a sign on the hat could be used to attract the public on a crowded thoroughfare, "the novelty of its apparent self-movement calling attention to the hat and its placard." (Figure 3.2)

In a day when women smoked furtively, the vanity with a concealed compartment for cigarettes must have had wide appeal. As patented in 1923 by Samuel S. Aber of New York, it had the usual space for cosmetics and, on the underside of the lid, a sliding mirror hid the cigarettes. If milady preferred, she could use the secret hideaway for money.

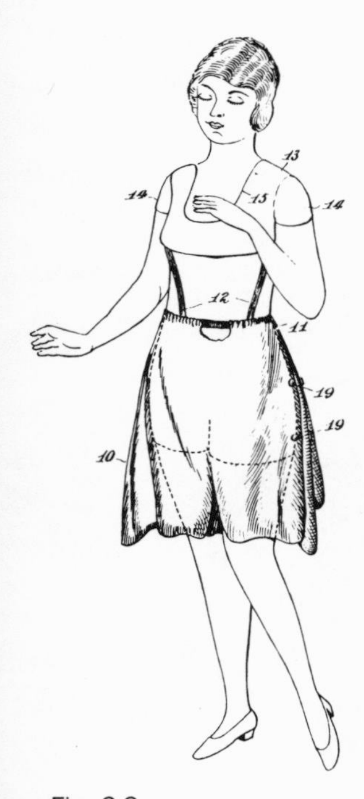

Fig. 3.3:
A 1927 bather covers herself.
(Patent 1,613,688)

Even in the mid-1920s, authorities frowned on display of the feminine body. As David Weisz of New York noted in his 1927 patent, many towns along the coast had ordinances compelling women bathers to wear outer garments between beach and home or hotel. His solution was a combined umbrella cover and garment. When she left the shore, the owner was to take it out of the umbrella and cover her shoulders and breast with the upper part—from which a modest skirt was suspended—concealing her from hip to knee. (Figure 3.3)

A more recent innovation is a double topcoat, especially useful at sports events, which a man can expand to enfold his girl friend if a storm arrives. Howard C. Ross of Gainesville, Virginia, an airline pilot, patented the garment in 1953. At the first drop of rain, the owner pulls his arm out of one sleeve, unsnaps the folded panels at front and back and invites his guest inside. She has the use of the other sleeve. (Figure 3.4)

A bottle for christening ships was conceived by William C. Loughran of Brielle, New Jersey with the fair sex particularly in mind. He said in his 1962 patent that breaking a thick champagne bottle, in the few strategic seconds before the ship slipped beyond reach, posed a real problem. "This feat is especially difficult for a lady selected to do the honors," he added, "as it usually requires several full 'baseball bat' swings and sometimes even with the able assistance of a husky escort it has taken ten such swings. . . ." He put on the market a bottle that fractured easily and liberated a fizzy chemical foam.

Inventors have produced many aids to feminine beautification. A striking example, patented by Martin Goetze of Berlin, Germany in 1896 is a device for producing dimples. The instrument looks like a brace and bit. The knob or bit—which is to be made of ivory, marble, celluloid or India rubber—is pressed on the site selected for the dimple and a massaging cylinder revolves around it as the handle is turned. (Figure 3.5)

Today's hollow-cheeked look, noted in stage and screen stars and fashion models, evidently was not popular in 1902 when Thomas C. Best of Chicago patented his cheek pads. He said they were to be held within the mouth to distend the cheeks and remove the appearance of emaciation. Each of the pair fitted between cheek and jaw, with a section under the upper lip, and had India rubber air chambers that could be blown up to the desired thickness.

Ways have been sought to avoid facial senescence without resort to surgery. Abbie M. Hess and Alfred Lee Tibbals of Kansas City, Kansas were granted a patent in 1913 for a means of removing wrinkles and double chin by pulling upward and rearward on the ears. A cord or elastic band, with metal ends that fitted in the ears, was stretched taut across the crown of the head.

Adolph M. Brown of Chicago exerted skin traction in a different way. He devised a series of little

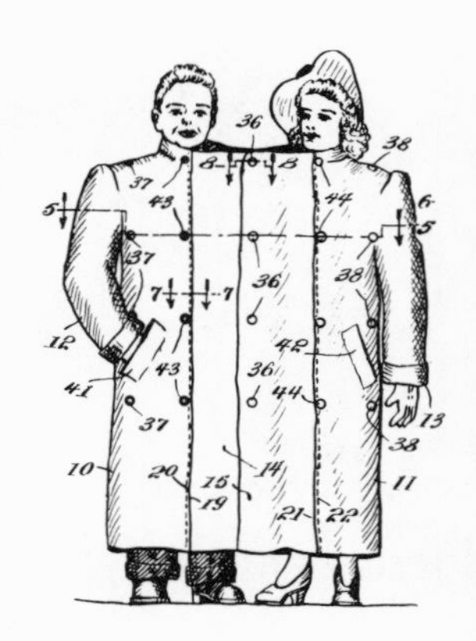

Fig. 3.4:
When it's cold ask her inside.
(Patent 2,636,176)
?a024

Fig. 3.5:
Brace and bit for making dimples.
(Patent 560,351)

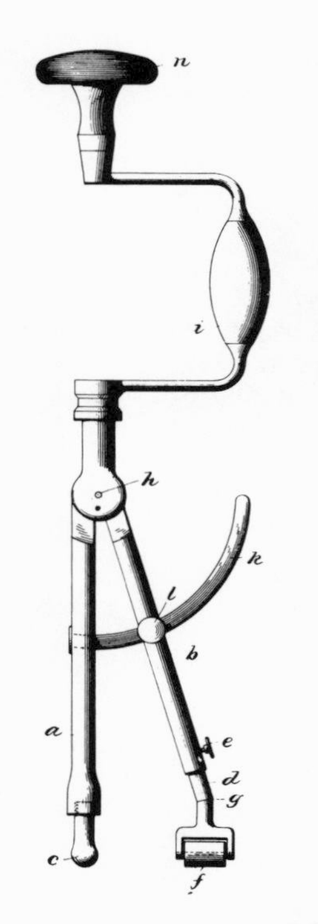

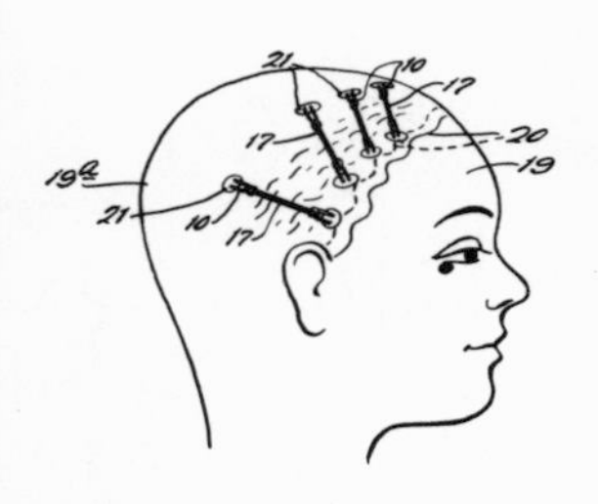

Fig. 3.6:
Scalp anchors pull away wrinkles.
(Patent 2,619,084)

anchors, to be inserted in the scalp and joined by rubber bands, so as to draw the skin taut over the face and hold it for several hours. The patent he received in 1952 advised coating the area over the anchors with a liquid that would give protection as it dried and allow the pull to be exerted for a considerable period. (Figure 3.6)

Two Florida men invented a lip printer that would enable a woman to paint her lips in a design especially made for her or copied from the mouth of a celebrated actress or beauty contest winner. Benjamin N. Greene and Sid S. Franklin of Miami Beach described in their 1953 patent a device obtainable at a cosmetics counter that would reproduce the desired lips. After coating it with lip rouge, the user was to press the frame on her mouth, leaving an outline to be filled in later with lipstick.

Mrs. Ann V. S. Mann of Petersburg, Virginia designed a beauty mask that a woman could wear at night without scaring her husband. In her 1954 patent she describes a flesh-colored plastic mask, painted and decorated on the outside with eyebrows, lashes and other attributes of a "strikingly beautiful woman." A wife wearing the mask during her hours of sleep could rest serene, "in the assurance given by her mirror that far from appearing grotesque, she is in reality a thing of beauty and that, actually, she sleeps in beauty." (Figure 3.7)

Fig. 3.7:
A beauty mask to wear at night.
(Patent 2,671,446)

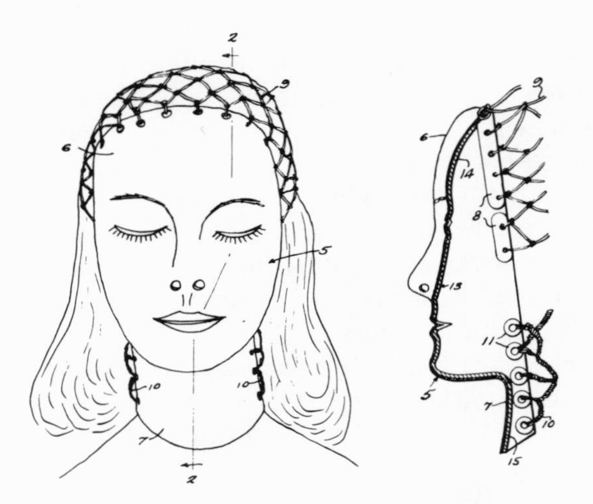

Fig. 3.8:
Hat that rests on the shoulders.
(Patent 1,045,060)

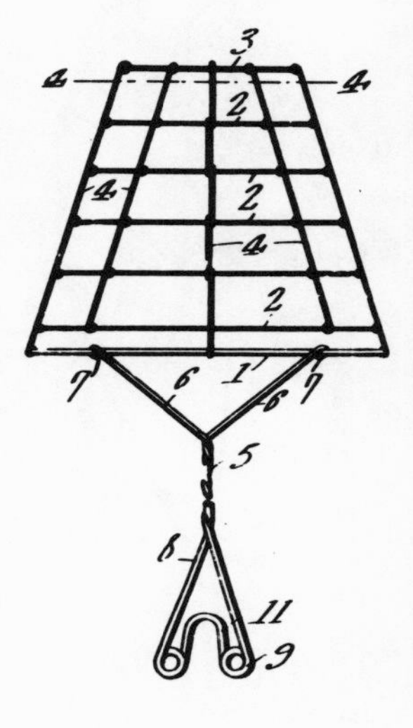

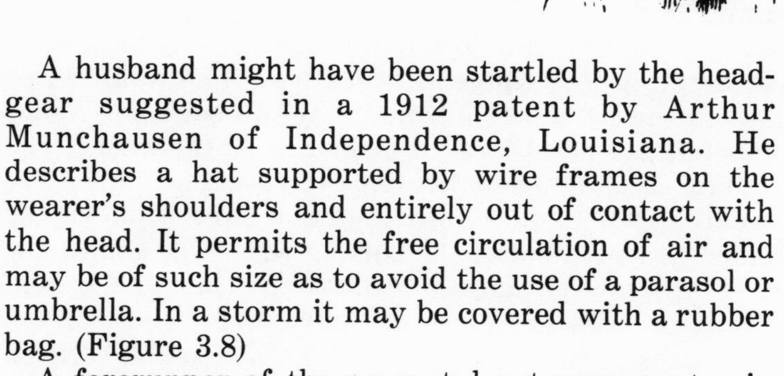

A husband might have been startled by the headgear suggested in a 1912 patent by Arthur Munchausen of Independence, Louisiana. He describes a hat supported by wire frames on the wearer's shoulders and entirely out of contact with the head. It permits the free circulation of air and may be of such size as to avoid the use of a parasol or umbrella. In a storm it may be covered with a rubber bag. (Figure 3.8)

A forerunner of the present-day transparent rain bonnet that covers the face as well as the hair was the lady's monocle hat patented in 1917. Joseph Stronge of St. Paul, Minnesota said his object was to protect the brows and eyes of the wearer, while permitting a look ahead and providing a novel effect. She could see with one eye through the glass or celluloid monocle set in the hat, while shielded from the observation of passersby. (Figure 3.9)

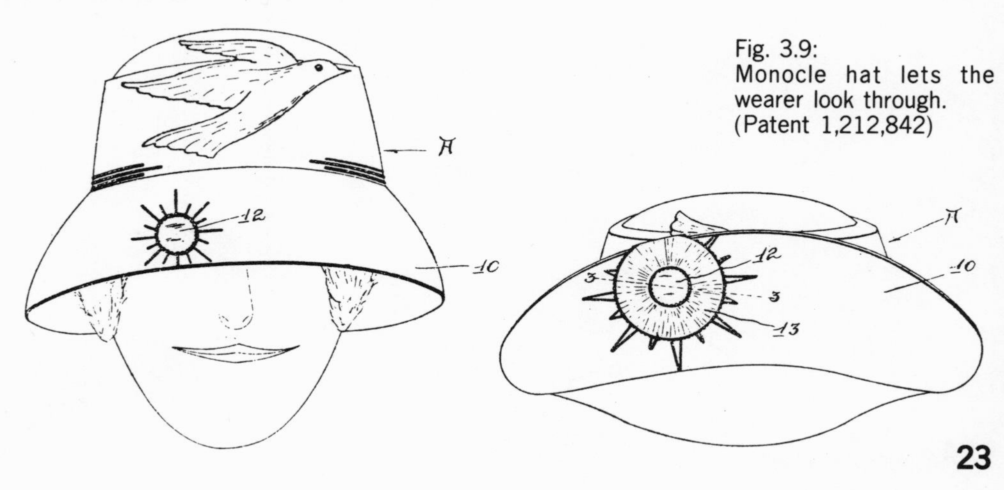

Fig. 3.9:
Monocle hat lets the wearer look through.
(Patent 1,212,842)

Fig. 3.10:
Baby hammock fits around the corset.
(Patent 484,065)

A quarter century earlier, in 1892, Charles C. Taylor of New York anticipated the present-day back carriers for babies. His patent describes a suspensory garment to be worn around a woman's figure. A graceful harness about the shoulders and waist held a small hammock "to carry babies, parcels and the like with little fatigue." (Figure 3.10)

Over the years other head-and-shoulder equipment has been offered for women's convenience. In 1905 Emmie Alice Thayer and Emily Waitee Thayer of Bellows Falls, Vermont patented a mirror to be hung in front of the face. "Ordinarily," they said, "a lady is obliged to hold a handglass in one hand while attempting to arrange her coiffure before a mirror or see to the fit of her garments at the back and hence has but one hand free for attending to the work." The mirror is supported by two wires hooked over the wearer's ears and a third movable up and down the chest. Adjustment of the bottom wire will "enable the user to view either the topmost feather of her hat, collar or even her skirt." (Figure 3.11)

A back-hair viewer, devised by Edward J. O'Brien of San Francisco, consists of a harness holding three small mirrors and an electric light, all suspended from the wearer's back, at about neck height. In his 1952 patent, he said his primary object was "to provide a portable rearview reflecting mirror assembly which may be conveniently worn upon the

Fig. 3.11:
Hang a mirror from your ears.
(Patent 790,537)

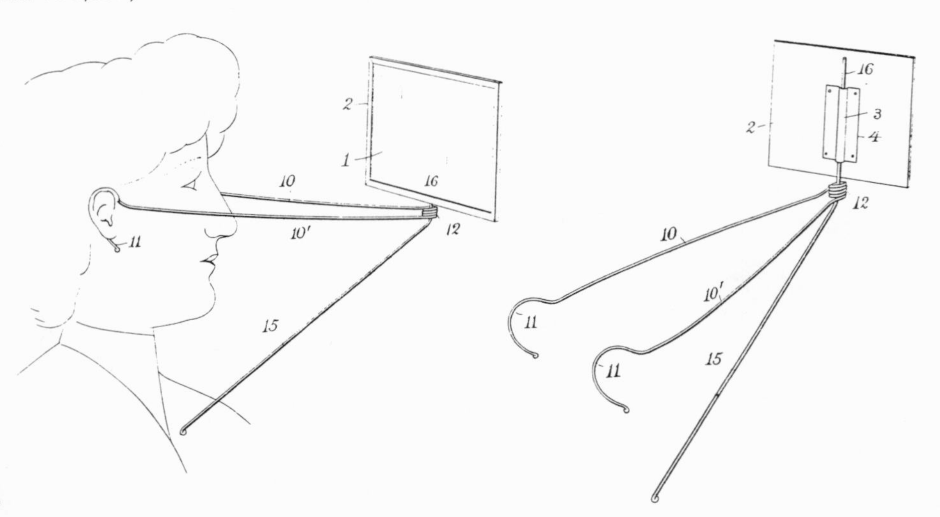

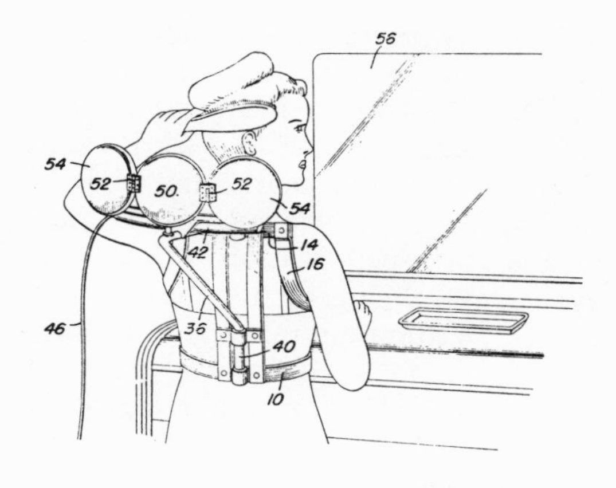

Fig. 3.12:
A triple view of the back of your hair.
(Patent 2,598,291)

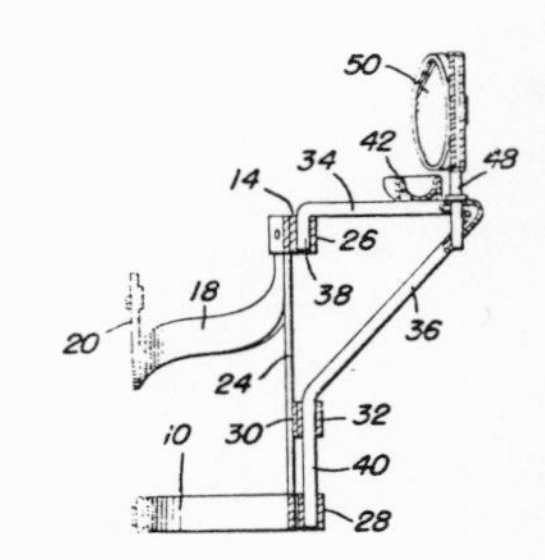

torso of the person for enabling examination of the rear of the head and brushing of the hair, while permitting unhampered use of both arms." The lady walked up to a large wall mirror, plugged in the lamp, and cast forward a full reflection of her back-hair. (Figure 3.12)

In that year women were having trouble zipping and unzipping the backs of their dresses. Roxann Shakarian Vahan of North Hollywood, California asserted that "in many instances this fastener extends from below the small of the back up to and between the shoulder blades or to the base of the neck and—the human anatomy being what it is—it is extremely difficult for the wearer to move the slide of the fastener between these two limits." The solution she offers is a tackle arrangement for the dressing-room wall. The lady is to back up to the frame, hook the zipper slide and move it up or down by pulling on one of two cords.

A posture-training earring warns the wearer by a bell-like sound when her angle is incorrect. As patented in 1952 by Peter Badovinac of Cleveland, Ohio the hollow piece of jewelry has suspended within it four little balls that tinkle against the sides and a fifth that rolls along the bottom whenever the earring is tilted out of the vertical.

Several means of self-protection have been offered the ladies—even before these dangerous days of mugging. For one who wants to know whether she is being followed there are the rear vision eyeglasses patented in 1956 by John Charles Reed of Chandler,

Arizona. A tiny mirror attached to each of the regular lenses can be adjusted for the desired view. Mrs. Natalie A. Stolp of Philadelphia patented a pointed defense against mashers in 1914. Rude and flirtatious youths and men, she said, "frequently avail themselves of the crowded condition of cars and other means of transportation to annoy and insult ladies next whom they may happen to be seated by pressing a knee or thigh against the adjacent knee or thigh of their feminine neighbor." Her patent discloses a spring to be attached to an underskirt, with a sharp point that protrudes in response to pressure.

What looks like a wedding ring, patented in 1972 by Charles Petrosky, an Arlington, Virginia barber, can be used by a woman to scratch an assailant for identification by the police. The weapon, to be worn on the ring or middle finger, has on the palm side a sharp little blade. It can rip the clothing or skin of an attacker.

Since the first patent law was passed in 1790, most American inventors have been men, and it is natural that patents have illustrated a gallant and protective attitude toward the gentler sex. More sophisticated inventions than those mentioned could be cited, such as the apparatus patented in 1968 by Rafael Carrera of Bethesda, Maryland for training married couples to achieve maximum efficiency in sexual relations. No doubt the benevolent approach will continue despite the growing complexity of technology.

4

In the Animal Kingdom
The Value of a Well-placed Tickle

The American inventor displays deep concern, not only for women and children but for his dumb fellow creatures, both furry and feathered. This affectionate attitude may be illustrated by a toothbrush for dogs, a tickle instead of a whip for balky mules and a diaper for parakeets.

The dog toothbrush was patented in 1961 by Bird A. Eyer, a Seattle realtor. As a dog fancier and judge at shows, he had noticed that otherwise deserving animals missed blue ribbons because of their teeth. The patented brush has rows of alternating rubber points and rubber cylinders. The recessed ends of the cylinders hold the toothpaste as it is applied and massage the gums. For the benefit of a Seattle journalist, Mr. Eyer demonstrated the brush in the family bathroom, applying it to Satchmo, his wife's toy French poodle. (Figure 4.1)

Fig. 4.1:
Well-brushed teeth for the dog show.
(Patent 3,007,441)

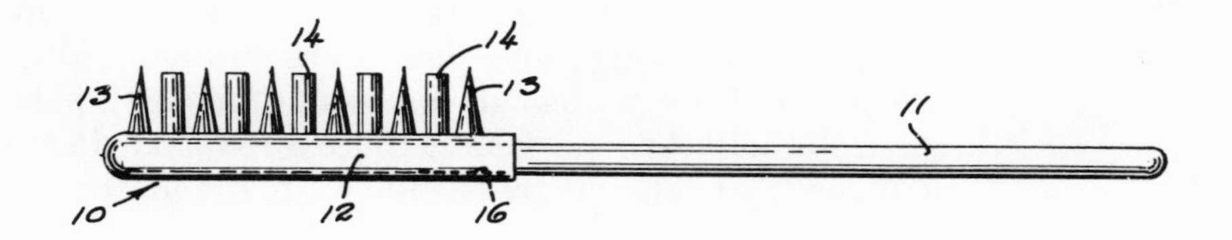

A doorbell for dogs is protected by a 1953 patent issued to Fred M. Adams of Shreveport, Louisiana. In what he calls an animal alarm scratching device, he combines a board on which the dog can sharpen its claws with a bell to alert the household. The invention is to be attached to the outside of a door, a favorite scratching place. It can also be installed inside the house, to give notice that the pet wants to get out.

William D. Davis of Redwood City and Albert J. Miller, Jr., of San Jose, California patented in 1963 a breath sweetener for dogs. Some mouth odors, they assert, detract from the owner's enjoyment of an animal's companionship. The preparation, in powder form, contains sodium bicarbonate and other compounds and may be sprayed from a squeeze bottle around the teeth. To make it palatable, the inventors suggest a flavor, such as spearmint, peppermint, wintergreen or rootbeer.

Among the canine beauty aids is an accessory for a veterinarian's clippers—a dog toenail groomer. According to the 1965 patent granted Charles Mueller of Oak Ridge, New Jersey the groomer beautifies but does not injure the animal.

The ears of boxers and other breeds are often cropped so that they will stand erect when the dog is mature. Florence Robertson and Carmine F. D'Amico of Valley Stream, New York got a patent in 1966 on a support for newly cropped ears, fitting atop the dog's head. It keeps other dogs from pulling at the freshly trimmed, sensitive lobes.

To make dogs more acceptable in the community, a Maryland patent attorney has devised a way to stop them from barking. He repeats their barks back at them, by radio but in a frequency inaudible to the human ear. John F. McClellan, Sr., of Monkton made a prototype of the instrument about twice the size of a transistor radio, and pointed it out of the window of his home. The neighbors' dogs were quieted and their owners did not notice. In discussing his 1972 patent, Mr. McClellan said, "I have discovered that dogs are subject to silencing by mimicry in much the same way that humans are."

Besides sound pollution, there are other ways in which dogs can damage the environment, indoors and

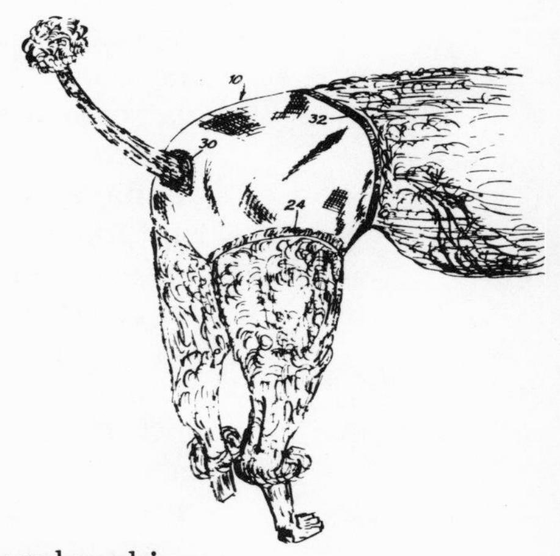

out. Training pants to be used in housebreaking which were patented in 1965 by David Hersh of Washington, D.C. are applied "to the posterior of the dog's abdomen for a period of time during which the dog is restrained from depositing its metabolic waste" on household furnishings. (Figure 4.2)

Fig. 4.2:
French poodle in training pants.
(Patent 3,211,132)

Fig. 4.3:
The Poop-Scoop for cleaner streets.
(Patent 3,685,088)

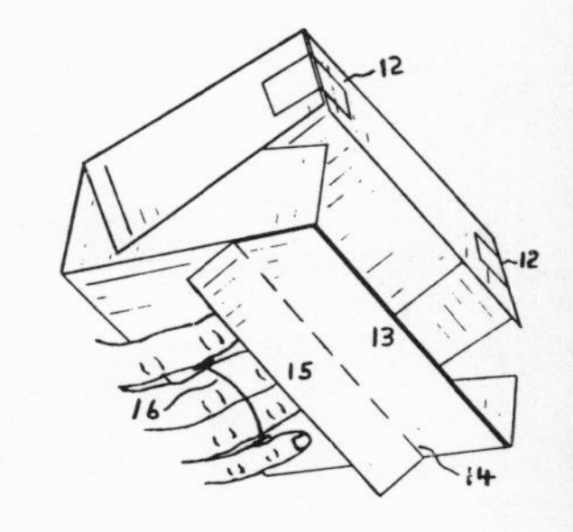

Taking pity on New Yorkers—whose streets had long been defiled—Henry Doherty of Wayne, N.J. patented in 1972 what he called the Poop-Scoop, with the more formal title, "Means for Collecting a Dog's Excrement by the Dog's Owner or Walker." Folded from a flat piece of paperboard, it is said to look something like a book. The user scoops the dirt in through the front and closes the box. Mr. Doherty had the assistance of his own dog, a cocker spaniel named Schuyler, in testing the device. (Figure 4.3)

Urban concern with the dog-waste problem was growing and other scoop patents followed. Late in 1972 Seymour A. Lemier of New York disclosed one that the dog owner, without dirtying his hands, can slide along the ground and use to load the refuse into a plastic bag. The bag "can be closed with a twist tie and deposited in a waste receptacle, leaving a completely clean scoop ready for subsequent use."

Alfred C. Gatti of Brooklyn patented in 1973 a pair of scoops attached to a handle and normally held apart by springs with an open-mouthed plastic bag between them. The operator can lower the scoops around the refuse and close both them and the bag. A pair of plungers keeps the scoops together until the loaded bag is thrown away.

Both cats and dogs benefit from some inventions, such as the scratcher patented in 1971 by David W. Hayward of San Diego. When the pet steps onto a platform, a switch turns on an electric motor, causing an arm to move a brush up and down against the animal's side. The motor stops when the animal quits the platform. According to Mr. Hayward, with experience, an animal can learn to operate the machine.

For animals with paws, Anna C. Gamble of Blue Island, Illinois got a 1971 patent on a push-and-pull sounding toy. It is described as a hard, shiny cylinder with a ball that rolls inside. The toy is too big to be carried in the mouth and crushed but small enough to encourage biting and pushing.

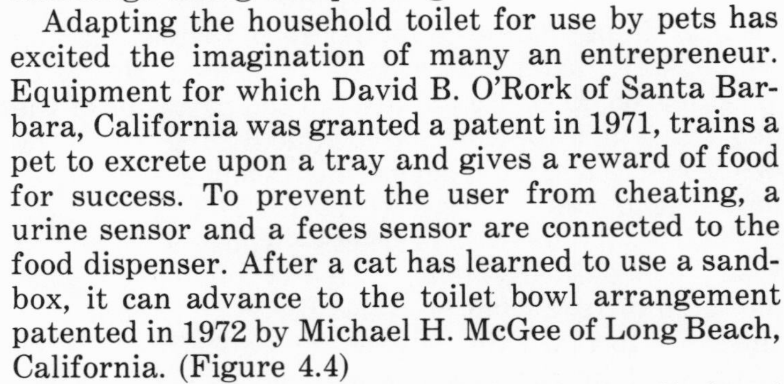

Adapting the household toilet for use by pets has excited the imagination of many an entrepreneur. Equipment for which David B. O'Rork of Santa Barbara, California was granted a patent in 1971, trains a pet to excrete upon a tray and gives a reward of food for success. To prevent the user from cheating, a urine sensor and a feces sensor are connected to the food dispenser. After a cat has learned to use a sandbox, it can advance to the toilet bowl arrangement patented in 1972 by Michael H. McGee of Long Beach, California. (Figure 4.4)

Fig. 4.4:
A sandbox graduate on the throne.
(Patent 3,688,742)

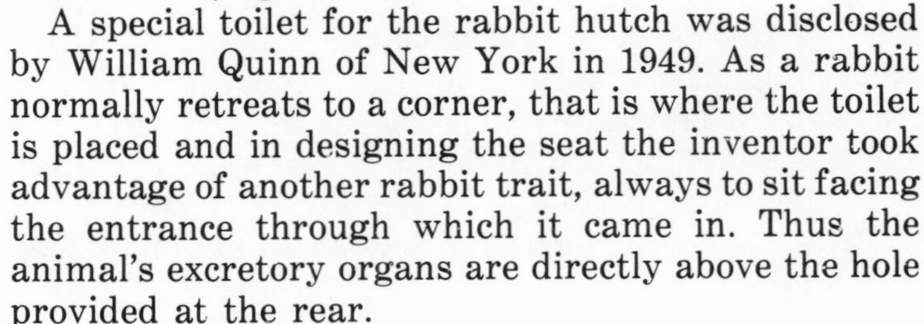

A special toilet for the rabbit hutch was disclosed by William Quinn of New York in 1949. As a rabbit normally retreats to a corner, that is where the toilet is placed and in designing the seat the inventor took advantage of another rabbit trait, always to sit facing the entrance through which it came in. Thus the animal's excretory organs are directly above the hole provided at the rear.

Equipment for cooling rabbits is offered by Alvin A. Ruport of Citrus Heights, California. As he explains in his 1953 patent, the animals, especially the does, suffer greatly in hot weather and may even fall fatally ill. Mr. Ruport installs on the floor of each hutch a flat coil of small-diameter metal tubing through which cool water from a hose circulates. The coil forms a reduced-temperature pallet or bed on which the rabbit can lie.

An early example of consideration extended toward feathered creatures is an eye-protector for chickens, invented by Andrew Jackson, Jr. of Munich,

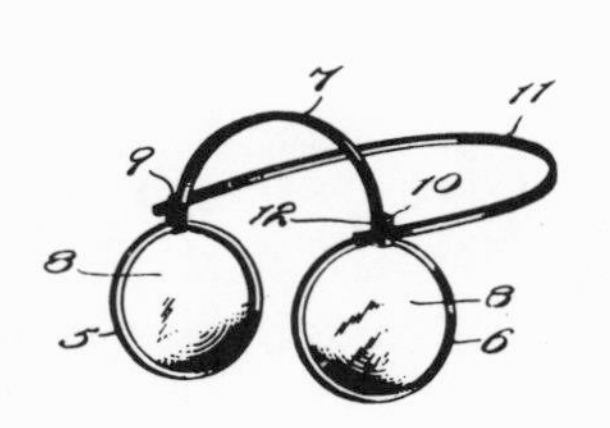

Fig. 4.5:
Glasses protect a chicken's eyes.
(Patent 730,918)

Fig. 4.6:
Fencepost holds nests and drinking water.
(Patent 947,929)

Tennessee. Drawings in his 1903 patent show what look like a pair of spectacles, whose frame is supported by wires going over the chicken's head and around the back of its neck. The lenses are of glass or mica and are not designed to improve vision but to protect the wearer's eyes from pecks by other fowls. The patentee's relationship, if any, to the seventh President of the United States is not indicated. (Figure 4.5)

A concrete fencepost patented in 1910 by L. Andrew Nelson of Foss, Wisconsin was only partly inspired by sentiment for the birds. He explained that one of his objects was

> "gaining for the toilers on treeless fields the inspiring, uplifting and encouraging companionship of the singing 'angels of the air,' by providing a fencepost which serves as a refuge, harbor, home and drinking fountain for a family of birds. The fenceposts studding the boundaries of a field in large numbers will, accordingly, furnish a large number of homes for the birds and in a large measure take the place of boundary trees. . . ."

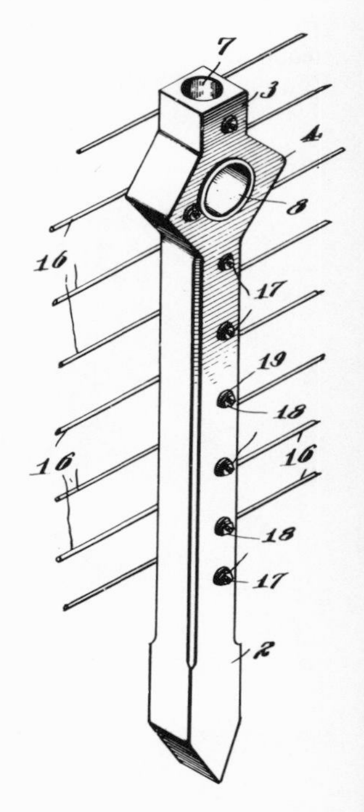

Mr. Nelson inserted tin cans to serve as nests and at the top of each post a cavity to catch drinking water for the birds. (Figure 4.6)

Many years later Bertha A. Dlugi of Milwaukee, Wisconsin attracted attention with her diapers for parakeets, birds that are usually allowed to fly about the house and frequently make deposits on the fur-

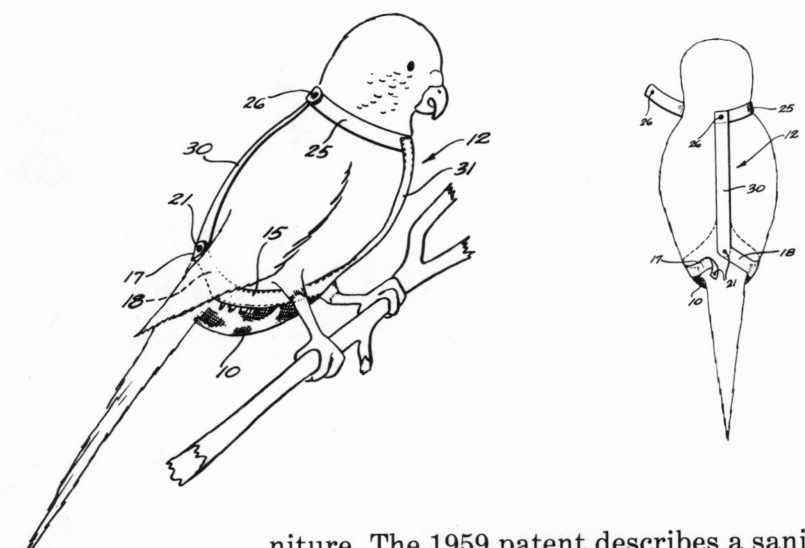

Fig. 4.7:
A parakeet is dainty in diapers.
(Patent 2,882,858)

niture. The 1959 patent describes a sanitary garment with a collar and tapes to hold an isosceles triangle of absorbent cloth underneath the tail. It is said that when the harness is put in place the bird will invariably ruffle its feathers, covering the strips of tape and making the garment inconspicuous. (Figure 4.7)

Two Floridians, Peter J. Scray and Cary Crews of Jacksonville, provide a caged bird with both companionship and fumigation. They patented in 1963 a hollow artificial bird to be placed alongside the live one on the perch. As the live pet swings them both, the fumes of a disinfectant emanate from an opening in the dummy's underside.

A California bird fancier revealed in 1972 a revolving perch that entertains its occupant. The bird can watch its own reflection in an attached mirror as an electric motor turns the perch around. William J. Dulle of Long Beach reports, "It has been found that the combination of rotation and changing reflections in the mirror is very amusing to the bird and therefore pleasing to its owner."

Chickens and other creatures can be trained to give public performances, sometimes with considerable financial gain for their trainers. Luis V. Millan Rohena of Santurce, Puerto Rico patented in 1967 a mechanical trainer for fighting cocks. One rooster is suspended in a body sling from a rod that swings around above a circular track, and the fowl being trained is encouraged to chase the captive. An elastic drill attached to the ceiling supplies the power. The bird being chased is covered with a transparent plastic sheet, which displays its plumage in natural colors, but protects it from being pecked. In any case, the trainee has no spurs.

A chicken plays basketball for food, fun and profit in an amusement apparatus patented in the same year by Grant Evans, technical director of Animal Behavior Enterprises, Inc., Hot Springs, Arkansas. When a spectator deposits a coin and pushes a slide, a table tennis ball is balanced on a column of air in front of the chicken. She pecks at the ball until she makes three baskets across the cage and collects her reward. The food rolls down a slide to within her reach. (Figure 4.8)

The patent also covers a version permitting a contest between the chicken and a human player. In this form, after the chicken pecks it, the ball rolls to the opposite side and is raised on a jet of air in front of a paddle gun. The human opponent manipulates the gun but is promised no reward.

Besides the chicken apparatus, Animal Behavior has rented out many other units along with the trained birds, rabbits, cats, pigs, raccoons, porpoises or whales to operate them. In slightly more than 20 years, the concern conditioned more than 7,000 animals of 40 different species. The renters have been tourist attractions, amusement and shopping centers, zoos and television stations.

Fig. 4.8:
Trained chicken plays a ball game.
(Patent 3,297,324)

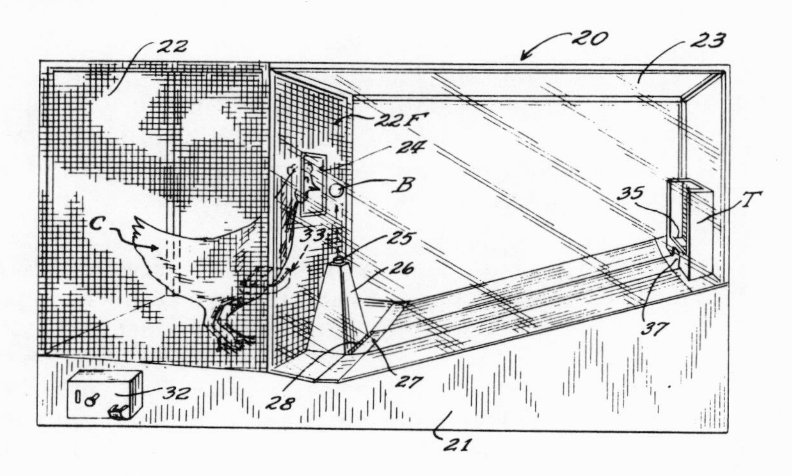

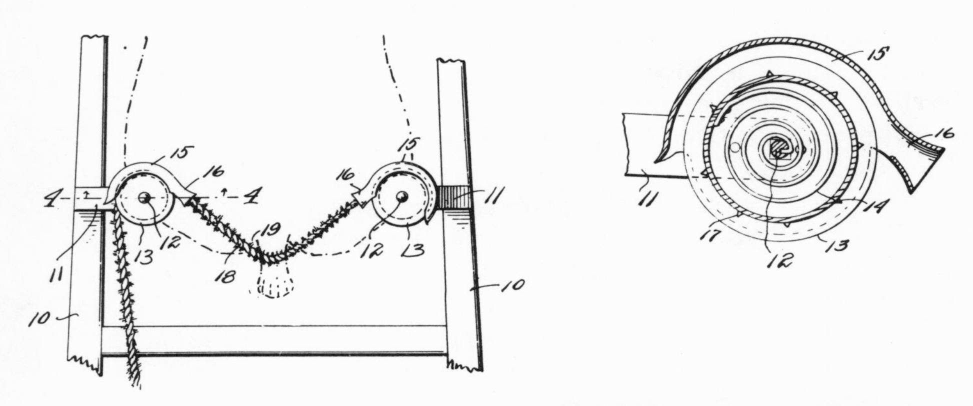

Fig. 4.9:
Tickle a mule under the tail.
(Patent 1,253,733)

Benevolent consideration is frequently bestowed upon large animals. Ilia Semotiuk of Vegreville, Alberta, Canada recognized in his 1918 patent that the mule is the most stubborn of all draft animals and is at all times inclined to balk. His answer, rather than the whip, is a tickling device consisting of a horsehair rope decorated with hog bristles. "Now it is well known by all students of natural history," he said, "that a mule's skin has little sensibility especially where covered with hair, as on most parts of the animal. It is also well known that the underside of a mule's tail at the part adjacent the body is without hair." The driver therefore is to saw the bristly rope back and forth under the tail to "so produce an irritating and tickling effect as to distract the attention of the mule and cause him to move forward." (Figure 4.9)

A cavalryman with long frontier service patented in 1857 a life preserver for horses. The inventor, Samuel P. Heintzelman of Newport Barracks, Campbell County, Kentucky made what he called a cavalry float consisting of a pair of bags, one for each side of the horse, connected with straps and inflatable with the breath. He said:

> During a long service on the frontier, where ferries are few, the necessity for some means more portable than boats for crossing the men and animals suggested itself. The want thereof often proves the source of much delay and loss, frequently the delay thus encountered in pursuing Indians being such as to make any further progress useless. Especially has this been experienced on the Pecos, in Texas, the Rio

Grande, the Colorado and its tributaries, and on the Columbia and its tributaries."

A squadron of cavalry so equipped, he added, if the men have waterproof pantaloons with feet, can cross rivers, lakes and estuaries dry-shod without the aid of boats. The horse life preservers, if big enough, would make it possible for an infantry soldier to ride behind each cavalryman and transport light artillery. The supports would also help emigrants cross rivers on their way to Oregon and California. (Figure 4.10)

As it happens, it was a Californian who many years later conceived another water convenience for horses—a swimming pool. A circular tank 40 or more feet in diameter, it provides about 10 feet of water and a ramp for access. Robert H. Shepard of Spring Valley, his 1966 patent makes clear, intends his pool especially for race horses, as swimming is an excellent conditioner for them. Controlling one or several animals by tether, a trainer walks around the rim.

Fig. 4.10:
Cavalry horse with 1857 life preserver.
(Patent 18,691)

A wiper-equipped windshield for a horse's rear is a novelty introduced by a 1937 patent. John E. Torbert, Jr., of National City, California says it is attached to the horse's tail. When the animal lifts its tail to push off flies, the wiper works. Safety reflectors also alert the driver of a following car at night, so that the horse will not be struck and injured. (Figure 4.11)

The Patent Office has been consulted about a wide variety of animals, wild and domestic. There is an anecdote about the Arkansas bullfrog farmer who brought in and exhibited a model of a feeder he had designed. It liberated flies and other bugs for bullfrogs three times a day. The inventor searched the prior art, but if he applied for a patent it was never issued. One patent that was issued went in 1971 to Shinichi Ishida and two other Japanese for an artificial reef in which fish could take refuge. And another, granted nine years earlier to John de Silva of Pembroke, Massachusetts unveiled a sleeping hammock for mink.

It may be appropriate to end this discussion of animal inventions with reference to a generally unfamiliar field—dental equipment for chinchillas. Frederick C. Fehrman of Washington, D.C. was concerned that chinchillas suffer malocclusion, or the slobbers, when their tusks grow too long. In 1953 he patented a small rotary grinder for the teeth and a device to hold open the animal's mouth. As a full-grown chinchilla weighs only a pound or a pound and a half, the mouth opener was about two inches long.

Fig. 4.11:
Windshield wiper on a horse's behind.
(Patent 2,079,053)

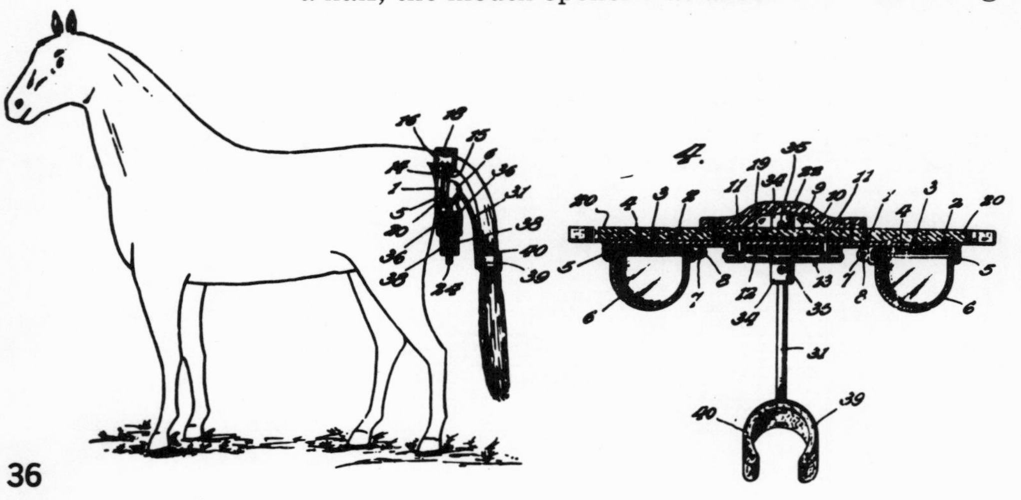

5

Sock and Buskin
See the Lion Chew Her in Two

Among the many inventors of things for the stage, the one who seems to deserve the palm is The Great Lafayette, described as a citizen of the United States and a resident of the borough of Brooklyn, city and state of New York. In 1907 he got a patent on an apparatus for staging an elaborate lion hunt, the capture of a lion and the feeding of a luckless female to the animal.

It is done with cages, false doors and real doors, a real lion and two fake ones. The scene opens in the jungle with a dummy lion concealed in a tree. When the hunters frighten it, it appears to jump (actually sliding on a wire) and lands in a trap. Then the live lion is dropped into the cage and displayed to the audience as if it had been the one that just jumped in.

In a series of moves, the real lion passes through a false door and is securely locked backstage, although the audience thinks it is about to devour a woman who has been pushed into part of the cage. The last the worried spectators see of the lion is its disappearing tail.

If this has been confusing, perhaps The Great Lafayette's explanation of what follows will clear things up:

> At the same instant that the real lion's tail disappears behind the false door, a man disguised in a

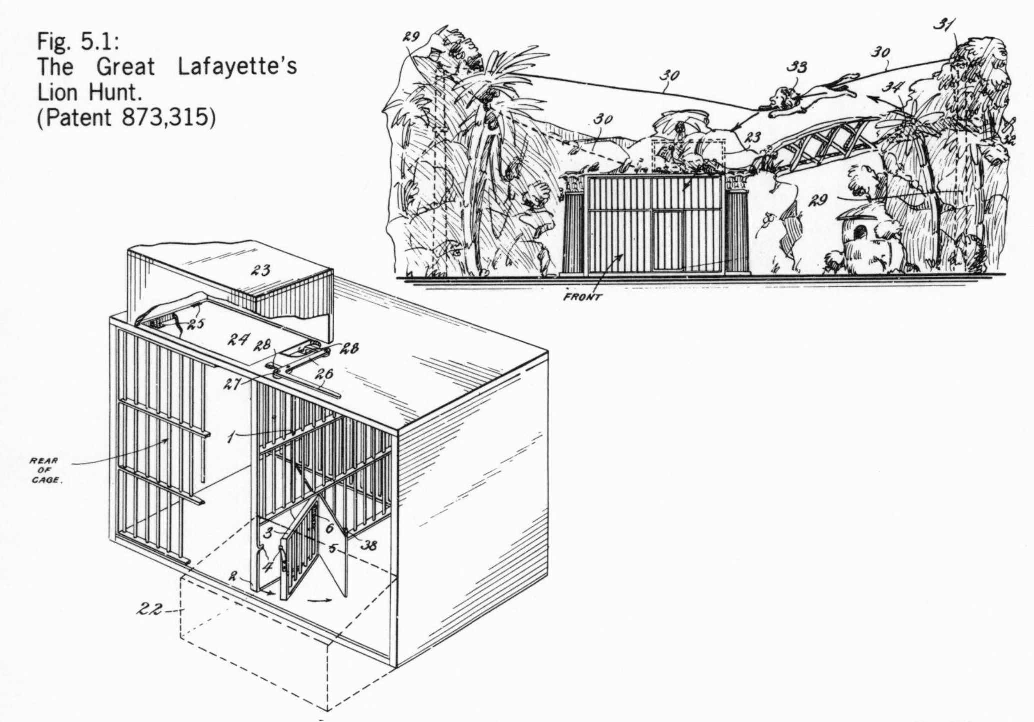

Fig. 5.1:
The Great Lafayette's Lion Hunt.
(Patent 873,315)

> lion's skin pushes through the opening controlled by the sliding door in the other side of the cage and emerges to the view of the audience. Thus as the real lion's tail disappears behind the false door, the sham lion's head appears to the audience on the other side of the false door, thereby creating the illusion that the real lion is passing through the opening in the partition to reach the woman. . . .The sham lion having pounced on the woman, the front of the cage falls out and then to reassure the audience the lion skin is thrown off and the trick revealed.

Simultaneously the audience is permitted to see the real lion, dinnerless, securely behind bars. (Figure 5.1)

The Great Lafayette (real name not disclosed) was not the first to devise stage equipment to fool an audience. A quarter century earlier, in 1882, James William Knell of Boston got a patent on his "Apparatus for Producing Illusory Dramatic Effects." He offered moving scenery in back and a horse trotting in place.

A screen of painted canvas, wrapped around a vertical post at each side of the stage, travels from left to

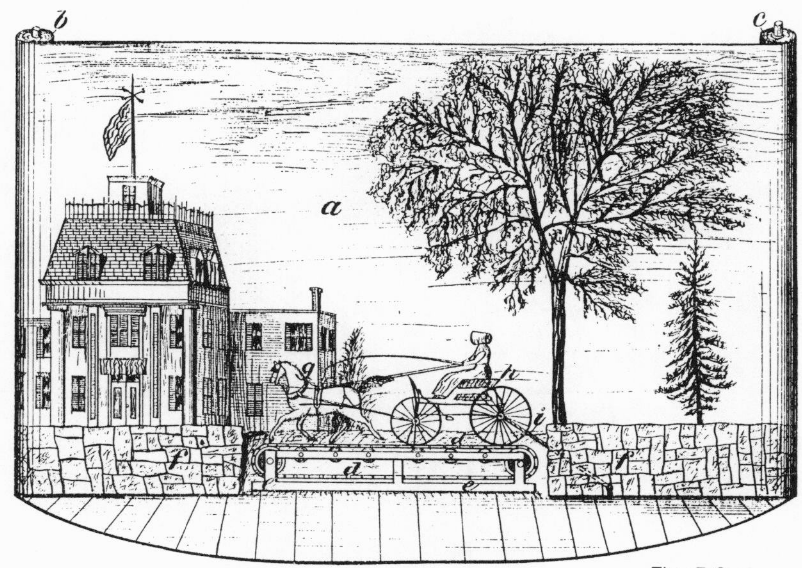

Fig. 5.2:
Scenery moves, and buggy stays still.
(Patent 256,007)

right displaying a panorama. Front and center is an endless belt that passes around two horizontal rollers and supports a horse hitched to a buggy. The horse and buggy are real and the driver can be too. As the horse paces toward the audience's left, it seems to be passing through the pictured countryside. (Figure 5.2)

There was little chance that a horse on a rolling belt would run away and endanger the viewers. But there has been growing concern since early in this century for audience safety, particularly in case of fire or panic. Henry Helbig, an American residing in Munich, Germany, saw protection in wide flights of stairs outdoors, onto which patrons could freely exit from any indoor level. In his day, real estate prices must have been relatively low, because the drawings in his 1906 patent indicate a frontage, including the stairs, three times the width of the theater.

Louis J. Duprey of Dorchester, Massachusetts went Mr. Helbig one better. His idea was to get the audience down and out. Under each seat, according to his 1924 patent, there was a trapdoor. In an emergency the theater management could touch a valve and lower all the seats simultaneously on their individual elevators. From the lower floor there would be easy escape to the open air.

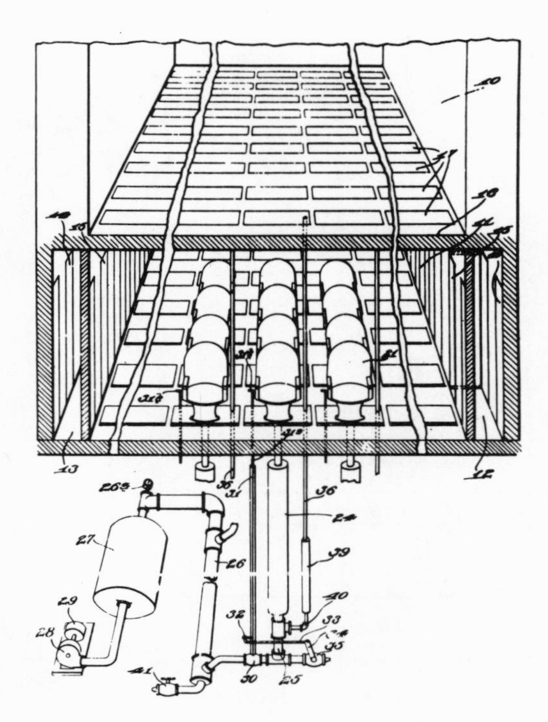

Fig. 5.3:
You leave the theater through the floor.
(Patent 1,517,774)

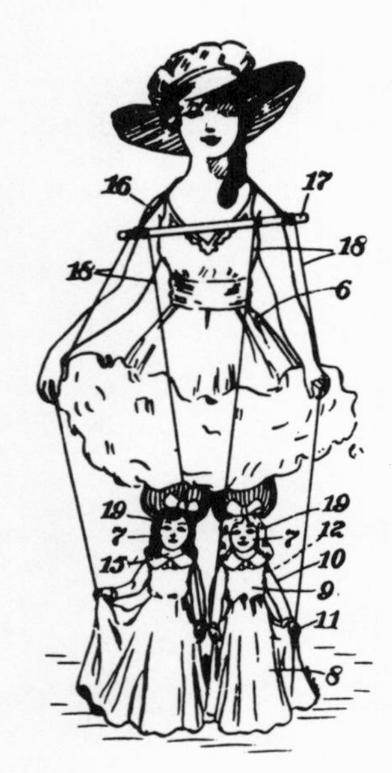

Fig. 5.4:
Puppets with knees for faces.
(Patent 1,505,942)

Arriving before the show, patrons were expected to enter from below, stopping on the way to hang their coats and hats on racks provided between floors. And if a man wanted to visit the washroom during the performance, he could go and come without treading on any toes. (Figure 5.3)

Elma Osborn Blanton of Jacksonville, Florida saw a use for toes on stage. In that same year, 1924, she disclosed a method of dressing a pair of legs as a pair of puppets, or miniature actors. The faces of the manikins were to be painted on the operator's knees, with wigs and hats fitted a little higher, like garters. Tiny bust pads could be placed at the proper level. A patent drawing shows a woman operator with two small dancers just below her bloomers. She is controlling the puppets' arms with strings.

The inventor points out that the operator should be relatively thin, so that the diameter of the leg just above the calf will represent a small puppet neck. "If the operators have any real muscular control over the knee cap" says the patent, "the facial expression of the manikin actors may be considerably changed at will." (Figure 5.4)

Lillian Russell, the operatic soprano, devised an off-stage convenience for the theatrical profession. It was a dresser trunk that she said answered the requirements of an actress in having all the cosmetics and necessities of make-up at hand, with mirrors and lighting fixtures arranged for quick use "as is necessary when the interval between acts is very short."

The patent drawings confirm Miss Russell's assertion that the trunk would present a neat and attractive appearance in the dressing room. When unfolded, it provided a table with drawers beneath and was designed to withstand the rough usage it would encounter on a theatrical tour. The patent was issued in 1912, the year of the famous actress's retirement, and bore her stage name instead of her real one—Helen Louise Leonard. (Figure 5.5)

Harry Houdini also invented something for life off-stage, although it was related to his specialty—extricating himself from handcuffs and sealed containers. Patented in 1921, it was a diver's suit design-

Fig. 5.5:
Lillian Russell's trunk turned dresser.
(Patent 1,014,853)

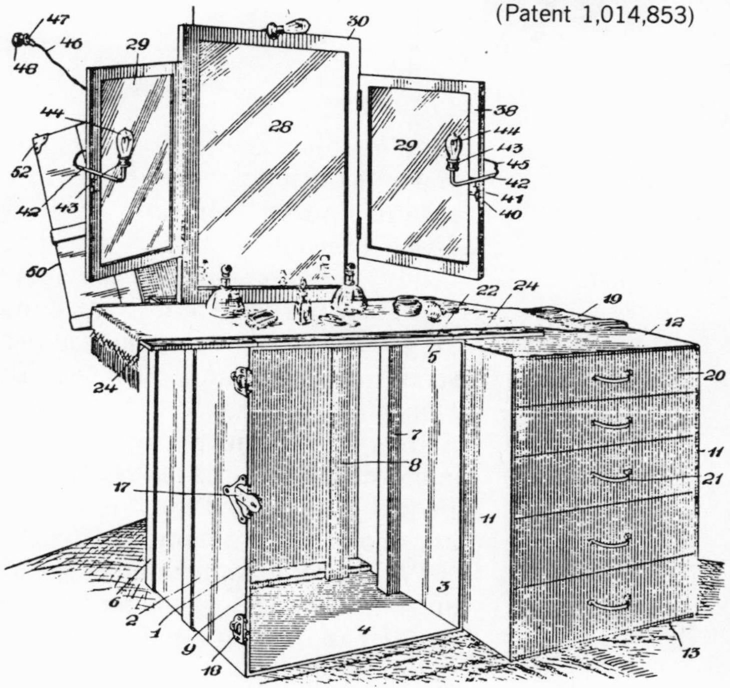

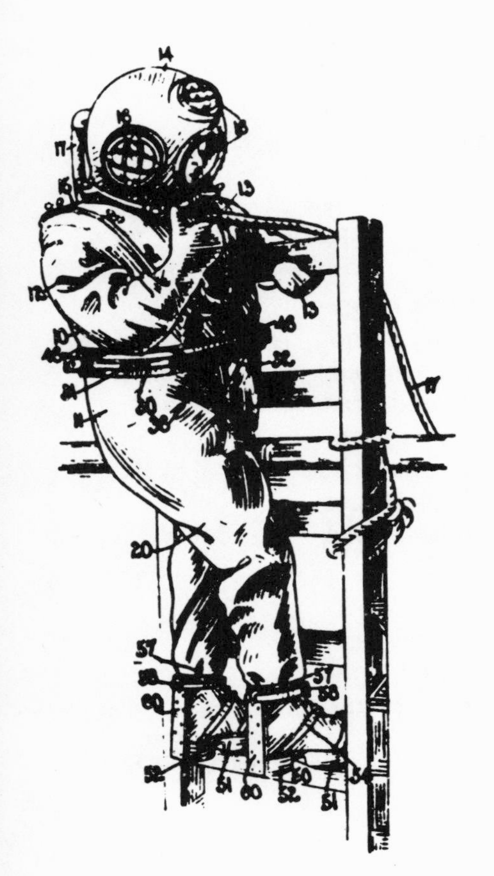

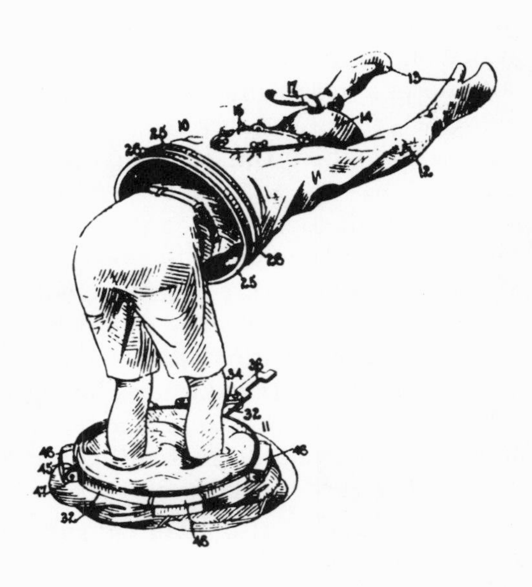

Fig. 5.6:
Harry Houdini's diver suit.
(Patent 1.370,316)

ed to enable the wearer to escape from it underwater. Harry said it was "arranged to permit the diver, in case of danger from any cause whatever, to quietly divest himself of the suit while being submerged and safely escape and reach the surface of the water."

As pictured, the suit divides like coat and trousers. To get out the diver opens the belt and lets the lower section drop around his feet. "By actual tests," the Houdini patent says, "it has been proved that not only a diver but an inexperienced person when submerged can escape from the suit in less than 45 seconds." The Brooklyn magician got the patent as Harry Houdini, not under his real name of Ehrich Weiss. (Figure 5.6)

Parking-lot amusements are sometimes based on simulated violence. Jennings William Carter, an American living in Manila, Philippine Islands patented in 1910 what might be called the "Kick-Me Game." A dummy in the uniform of a policeman or soldier is leaning on a fence and has a long neck that can slide upward. When a board at the seat of the figure's pants is kicked, the neck rises and the strength of the kick can be read on a graduated scale below the dummy's ears.

In serious terminology, Mr. Carter describes his invention as a testing machine that can be used to determine the force of a blow but he discloses that it may well be operated as part of a vending machine. A patron is permitted a kick for a coin. (Figure 5.7)

Another non-theater amusement device has a genuine human target, lying on his back with his legs in the air and his bottom fully exposed. A contestant is encouraged to stand on a small platform some distance away and throw "missiles" (pictured as baseballs) at the target. However, as planned by Anton Hulsmann of San Francisco in his 1923 patent, there is a control invisible to the spectators, operable by the target player, which shakes the platform on which the thrower is standing, knocking him off balance and deflecting his aim. The target can also move his body right or left. Mr. Hulsmann counts on both the grotesque position of the target and the confusion of the pitcher to produce hilarity and amusement among the viewers. (Figure 5.8)

Besides mechanical means for play, there can be mechanical aids for the playwright. For example, Arthur F. Blanchard of Cambridge, Massachusetts patented in 1916 what he called a movie-writer or scenario-forming device. This means of inspiring an

Fig. 5.7:
Kick-me game measures foot-power.
(Patent 968,325)

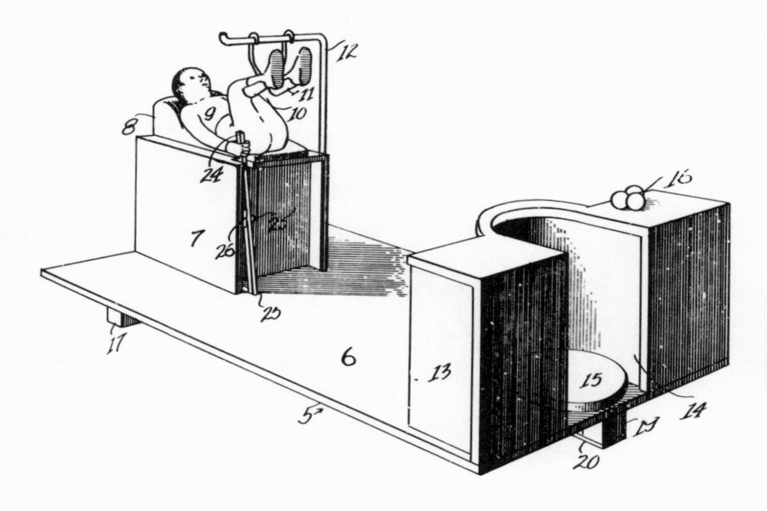

Fig. 5.8:
Human target with secret defense.
(Patent 1,467,934)

author is a box containing a half-dozen rolls of paper on which thought-stimulating words are inscribed. As the rolls turned, words were displayed through openings in the box. Pictures and bars of music could also be shown.

With his equipment, Mr. Blanchard asserted, original ideas and suggestions of fiction could be evolved without any effort on the part of the operator other than turning the rolls. As an example, the machine might turn up six words that could serve as the outline for a short story: aged, aviator, bribes, cannibal, carousal and escapes. "These particular words readily suggest, for instance," he said, "that an aged aviator after flying through the air on a long trip, lands finally on a desolate island where he is met by a cannibal, whom he is forced to bribe to secure his safety. After an interim which is full of possibilities . . . a carousal ensues following which the aviator escapes." The patentee remarked that his mechanical movie-writer might be used not only for literary work and instruction, but as a parlor game.

Those were the days of silent pictures for which phrases such as "I hate you" or "Drop that gun!" had to be lettered on the margin or on a separate screen. Borrowing the balloon technique from the comic strips, Charles F. Pidgin of Winthrop, Massachusetts arranged to display speech on the movie screen coming right out of the actors' mouths. His 1917 patent describes rubber balloons blown up by the players

Fig. 5.9:
How to make silent movies talk.
(Patent 1,240,774)

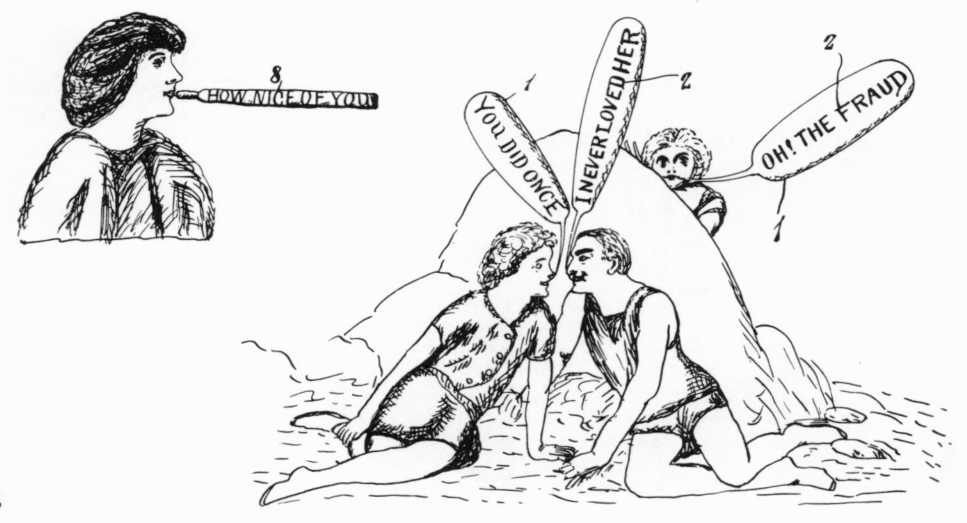

and displaying printed words on their sides. They could be gradually inflated, enlarging the visible sounds as the characters' emotions grew. Paper bags or spirally coiled tubes were other means to carry the lettering. But only the talk would show on the screen; the outlines of the bags, balloons or tubes were to be neatly erased in developing the film. (Figure 5.9)

Long afterward, when children knew only movies with sound, a Boston man conceived a theater that would offer them full audience participation. They would be able to gallop after the bad men, as members of the sheriff's posse, whooping and shooting cap pistols. Frank E. Leahan's 1955 patent shows, in place of theater seats, a troop of mechanical horses heading toward the movie screen. The mounts bob up and down, a breeze blows against the riders' faces and ribbons of light moving from front to rear add to the illusion of forward speed. In the proposed theater, the young horsemen could not only watch characters on the screen ahead and listen to full sound effects—but see fellow members of their posse pictured on a second screen at the side of the auditorium. (Figure 5.10)

Fig. 5.10:
Better than sitting and watching.
(Patent 2,719,715)

Fig. 5.11:
Walt Disney's Mad Teacups.
(Patent D.180,585)

Fig. 5.12:
Edgar Bergen's smiling doll head.
(Patent D.129,255)

Mad Teacups, right out of Alice in Wonderland, already existed in Disneyland, near Los Angeles, when Walt Disney got a design patent for them in 1957. Eighteen of the cups—really cars big enough to hold four or five children—were arranged around a turntable in groups of six. The turntable, which might be called a tea tray in the sky, revolved and each circle of six cups revolved. A cup—designated in the patent "a passenger-carrying amusement device"—could be entered through a hole in the side. By holding a handgrip that was mounted in the center like a table, a cupful of children could swing their cup around in either direction. (Figure 5.11)

As is natural, a good many stage personalities have produced inventions in their fields. Herb Shriner (more formally Herbert Schriner) of Larchmont, New York, a comedian and professional player of the mouth organ, was co-patentee in 1960 of a combination harmonica and water pistol called the Harmonigun. Other entertainers are credited with things that have little or nothing to do with the theater. On the borderline are doll heads for which Edgar Bergen, the ventriloquist, received two design patents in 1941. (Figure 5.12) Natacha Rambova (also known as Winifred Guglielmi and Mrs. Rudolf Valentino) patented in 1912 a doll that could serve as a wrap. Hedy Kiesler Markey (stage name Hedy Lamarr) and George Antheil, a composer, got a 1942 patent on a secret communication system for torpedo control. In 1950 and 1951 Lawrence Welk was listed as designer of two lunch boxes. More recently Edie Adams (named on a 1965 patent as Edith A. Kovacs and in 1966 as Edith Adams Mills) protected, through patents, her ring-shaped cigarette and cigar holders. And in 1969

Herbert Zeppo Marx was co-inventor in two patents for a heart alarm to be worn on the wrist.

From a complete outsider, Darrell M. Johnson, an automobile dealer and former mayor of Thomson, Georgia comes something that might well appeal to the Sock and Buskin set. To aid public speakers, he created a mechanical listener whose eyes light up and whose eyelids flutter at the sound of the human voice. Patented in 1960, his mechanical captive audience was formally named "a device for stimulating mental processes."

The idea struck Mr. Johnson as he was preparing a speech for a sales meeting. The patent pictures several human embodiments: sculptured heads, plaster reliefs, and a seated man. A hidden microphone switches on lights behind the eyeballs and animates the eyelids. Mr. Johnson thought that his listener would help, among others, politicians, especially if they suffered from mike fright or television stiffness. It might help actors too. (Figure 5.13)

Fig. 5.13: Mechanical listener's eyes light up. (Patent 2,948,069)

6

From Here to There
Wings, Wheels and Paddles

A set of personal wings that was patented a century ago might well be transporting today's commuters and traveling men if it had caught the popular fancy. Watson F. Quinby of Wilmington, Delaware invented a flying apparatus to be worn by one person, with a dorsal wing on the back and a wing on each side, all to be flapped by the arms and legs.

Mr. Quinby, in his 1872 patent, compared some of the movements to those of winged animals and said the wing points resembled those of a bat. The entire weight, which need not exceed 15 pounds, was to be supported on a ring around the flyer's waist. In the inventor's words:

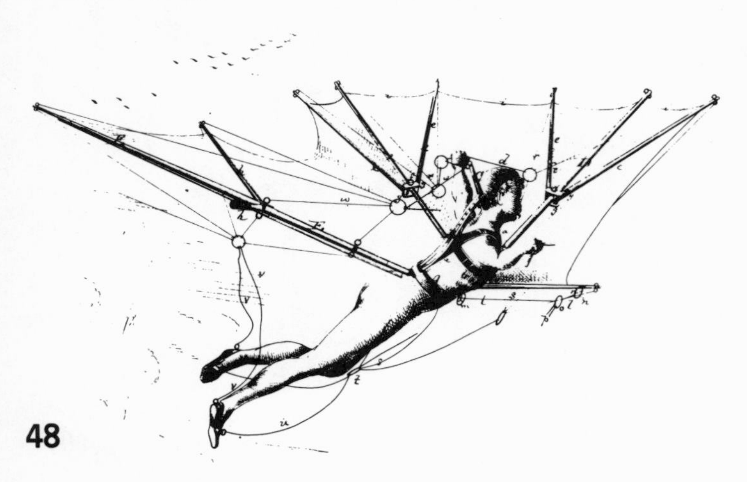

Fig. 6.1: Just as easy as swimming. (Patent 132,022)

> It is intended to start from the ground. In order to make a beginning one foot is disengaged from the stirrup when—by raising the other foot and pushing the hands upward and forward as in swimming—the wings are raised. Then, by suddenly depressing the wings, by means of the elevated leg, the former are intended to elevate the body by their action on the air. This alternate elevation and depression of the wings is continued as long as flight is desired. After rising from the ground the other foot may be inserted in its stirrup and both legs used. The actions are intended to be natural, resembling those of swimming in water. (Figure 6.1)

For some time thereafter, inventors studying means of flight seem to have concentrated on lighter-than-air craft. In 1887 Charles Richard Edouard Wulff of Paris made a notable contribution with his patent for a balloon that was guided and propelled by birds—including eagles, vultures and condors.

Previous attempts to guide or steer balloons, he said, had comprised mechanical, electric or other motors, and these were too heavy and unsuited to the medium. He substituted living motors and proposed arrangements by which "all the qualities and powers given by nature to these most perfect kinds of birds may be completely utilized."

M. Wulff said the balloon might be of any suitable form and dimensions. A car was suspended below it. "In the car," he added, "is placed the aeronaut, who has charge of the whole apparatus, and a reservoir of gas may also be placed therein to provide for leakages."

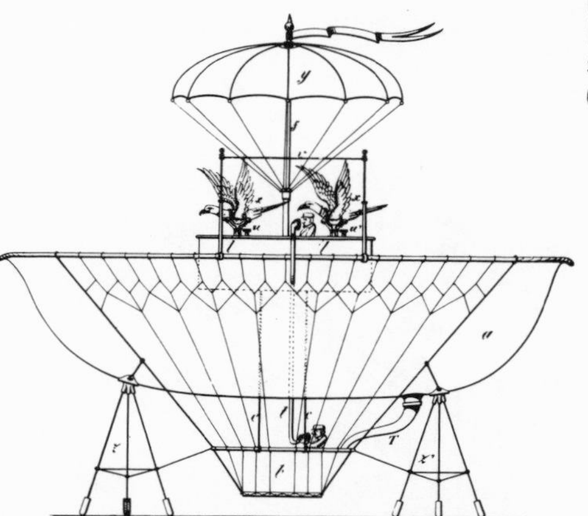

Fig. 6.2:
Birds for both power and steering.
(Patent 363,037)

On a floor above the balloon were an "overseer or person working the apparatus" (we might call him an engineer) and the birds. The aeronaut, from his position below, gave orders to the engineer through a speaking tube. Surmounting the craft was a parachute, to protect the birds and regulate the descent of the balloon.

Four birds were attached by bands and shoulder straps to a railing on the upper platform. The engineer could rotate the framework to head the birds in whatever direction he liked. Their "corsets or harnesses" could also be tipped so as to angle their flight upward or downward. "It may be observed," said M. Wulff, "that the birds have only to fly, the direction of their flight being changed by the conductor quite independently of their own will." (Figure 6.2)

Other early aircraft were propelled and to some extent lifted by manpower. A flying machine patented in 1889 by Nicholas H. Borgfeldt of Brooklyn has a cigar-shaped balloon from which is suspended a long, shallow car, like the skeleton of a ship with raised bow and stern. The inventor says his flying machine "shall be so balanced that the balloon connected therewith will be incapable of elevating the car and the contents thereof by its own buoyancy, the wings being required to cause the whole apparatus to rise." Six men standing on the car floor were to flap sails in

Fig. 6.3:
Flapping the wings of a flying machine.
(Patent 411,779)

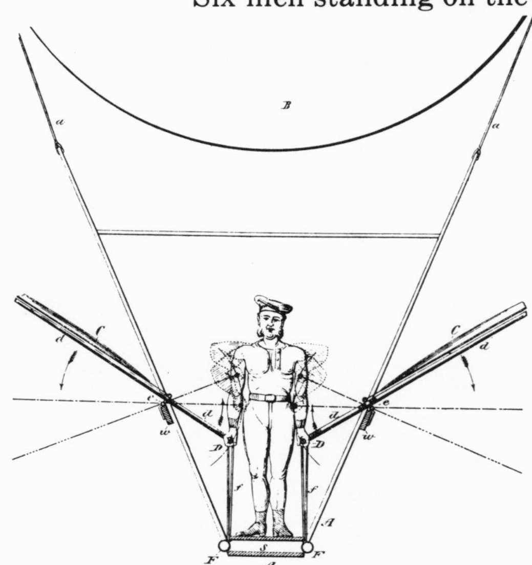

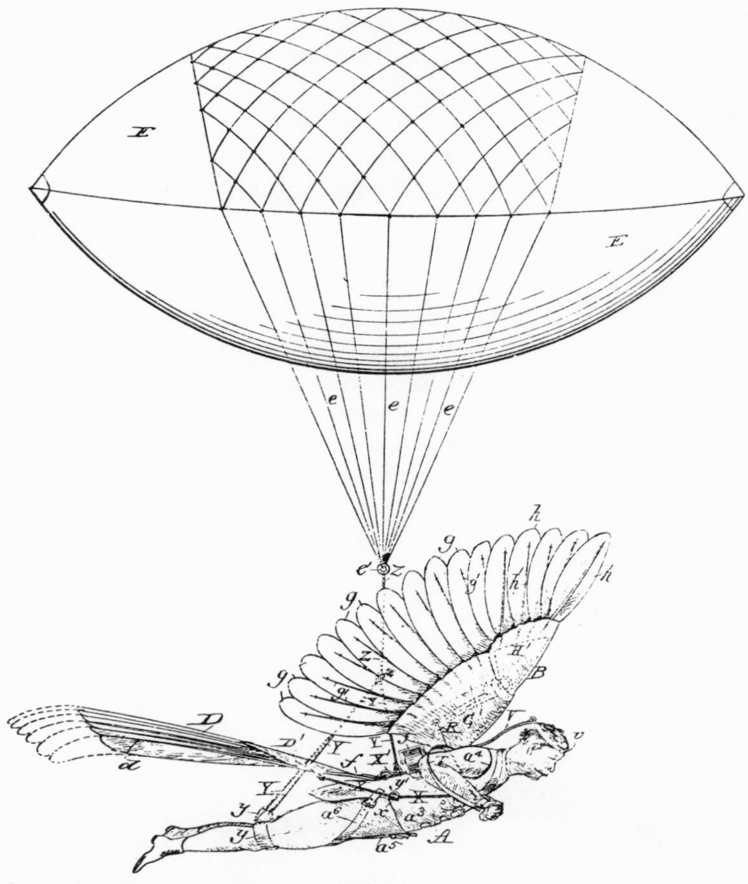

Fig. 6.4:
With a little balloon and a tail.
(Patent 398,984)

the wings to gain elevation. Floats would prevent the car from sinking if it struck the water. (Figure 6.3)

Another but smaller machine, patented in the same year, is intended for a single passenger, who wears a jacket with wings attached and sports a long tail. Reuben Jasper Spalding of Rosita, Colorado provides a small balloon for partial support and gives the flyer levers with which to flap down the two wings. They return upward under spring pressure. Artificial wingfeathers and tailfeathers offer proper surfaces to the air. Mr. Spalding says the wings operate with practically the same effect as those of an eagle, and by manipulating them and the tail, the aeronaut can change his course. (Figure 6.4)

Edwin Pynchon of Chicago offered a novel method of propulsion. His 1893 patent discloses an airship supported largely by gas and moved through the air by the explosion of powerful cartridges. Drawings picture what looks like a large marine vessel, displaying on its bow the name Albatross, with a buoyancy chamber on top, wings at the sides and smoke from high explosives coming from the stern. The elaborate equipment includes a bell to notify the engineer when a cartridge is in place, ready to be discharged. (Figure 6.5)

While innovative Americans were exploring travel through the air, others were working on the railroad. In 1868 Zadoc P. Dederick and Isaac Grass, both of Newark, New Jersey, patented a steam carriage in the shape of a man pulling a cart. He is shown smoking a pipe and marching forward wearing a stovepipe hat. The boiler is in the torso. The legs are jointed and move forward and backward, one at a time. A driver sitting in the cart behind can control the length of the steps, cause the feet to clear obstacles, and change the direction in which the figure is walking. The machine can also be caused to move backward in long or short steps. (Figure 6.6)

Fig. 6.5:
A ship literally shot through the air.
(Patent 508,753)

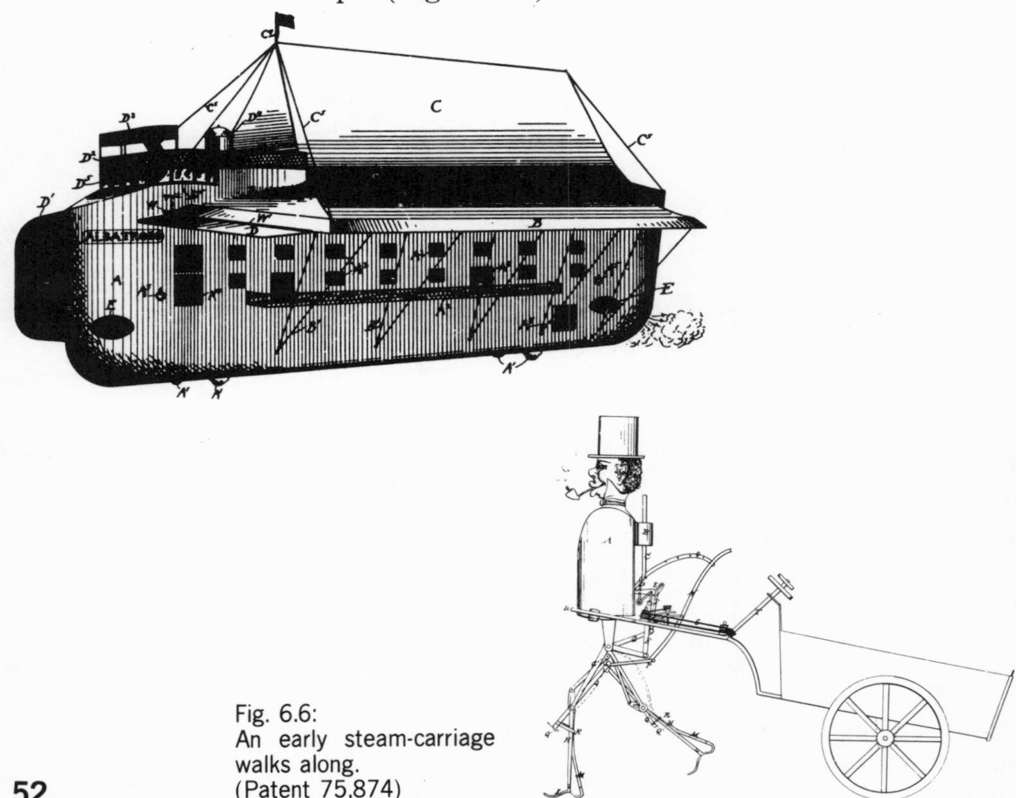

Fig. 6.6:
An early steam-carriage walks along.
(Patent 75,874)

Also trackless, a road locomotive patented in 1878 by John B. Root of New York had wheels that changed their angle periodically from the perpendicular and aimed diagonally at the ground, now right and now left. This was to avoid embedding themselves in the dirt of the roadway, a problem that afflicted tractors. In one form, the locomotive was propelled by an engine, and in another by treadles that the driver operated. A third form—without wheels—was intended for ice, but the means of propulsion is unclear. (Figure 6.7)

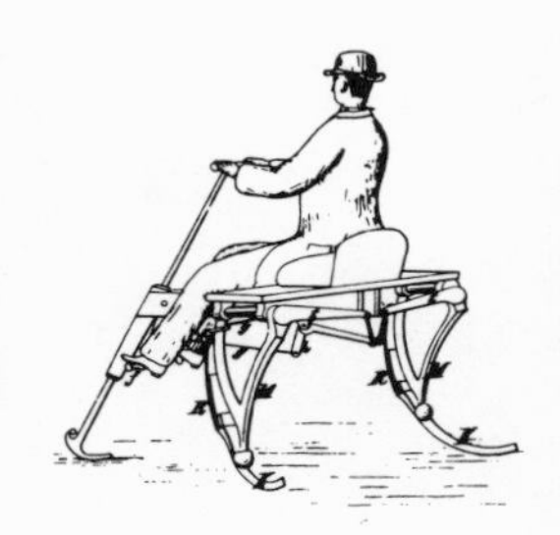

Fig. 6.7:
A treadle locomotive for land or ice.
(Patent 205,212)

With the advent of rails, engineers had trouble with animals and people at road crossings. William Bell of Verona, Mississippi got an 1885 patent on what he called lazy-tongs to be extended by the engineer in front of the locomotive. The engineer could also frighten animals with a steam whistle and spray them with water or steam from a nozzle, both installed at the front end of the folding tongs. (Figure 6.8)

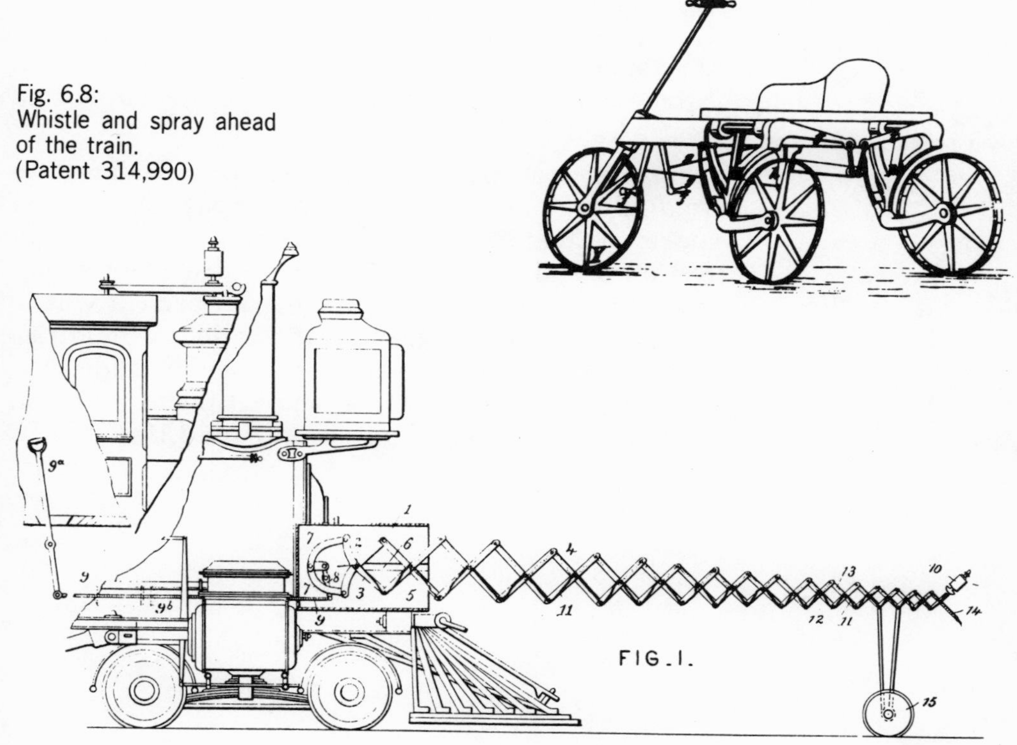

Fig. 6.8:
Whistle and spray ahead of the train.
(Patent 314,990)

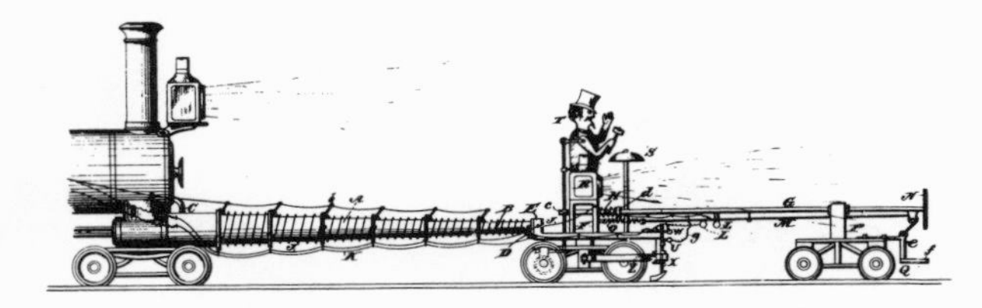

Fig. 6.9:
Bang, bang! Get off the track.
(Patent 386,403)

Jack William James of Cuba, Tennessee went Mr. Bell one better with the collision-preventer he patented in 1888. Attached to a telescopic tube ahead of the engine he placed a flatcar, on which was seated a dummy ringing a gong with each revolution of the wheels to frighten cattle. In front of the flatcar was a pole, supported by wheels. If the pole struck a train coming in the opposite direction, or some object like a log, a man's body or a cow, an electrical circuit applied air-brakes, reversed the engine and pulled in the pilot car. If there was an open bridge ahead, the pilot car would fall through and the train would be halted.

"By means of this pilot apparatus operated by compressed air, electricity and mechanism, as described," Mr. James concluded with modest understatement, "the danger of collisions and other accidents from obstructions, etc., on the road can be prevented to a great extent." (Figure 6.9)

Another transportation system was designed to avoid all ground hazards by moving high in the sky. Andrew J. Morrison of Buffalo, New York patented in 1885 an aerial railway to be suspended from a continuous cable supported by balloons. Each car was to be moved by its own gravity; the balloons could rise or lower to provide the proper downhill angle. Mr. Morrison also provided a gas compartment for the top of the car, to make it lighter. (Figure 6.10)

Fig. 6.10:
Aerial railway runs by gravity.
(Patent 328,899)

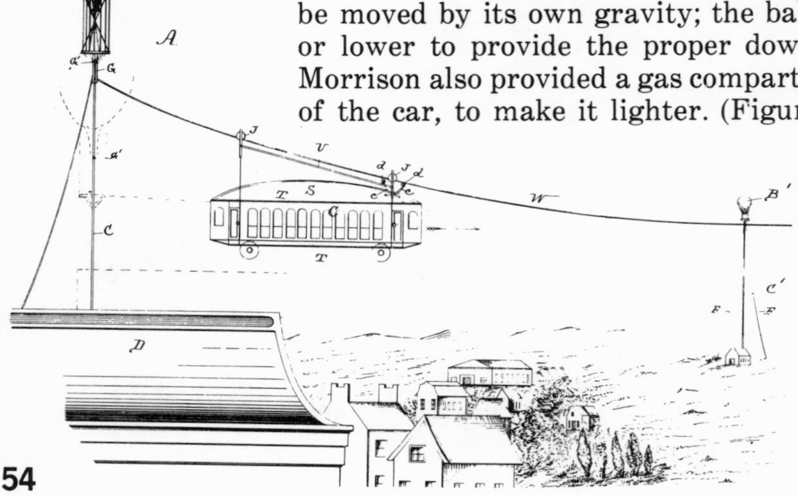

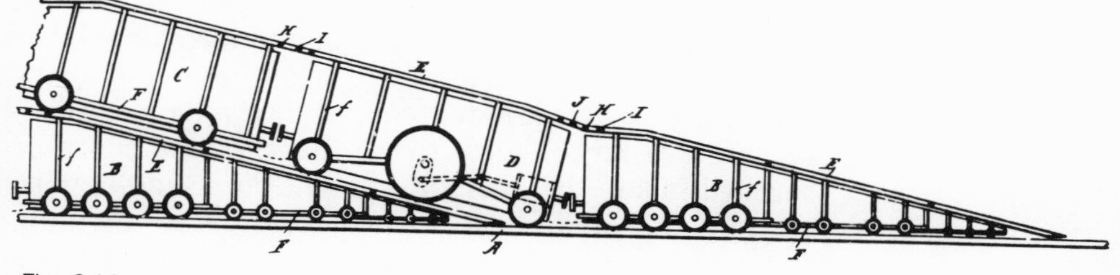

Fig. 6.11:
Train runs up and over another.
(Patent 536,360)

Ten years later, in 1895, a patent granted to Henry Latimer Simmons of Wickes, Montana portrayed a train that, when it met or overtook another, could run up over it and dismount again. The lower train was equipped overhead with rails that were of the same width as the regular track. Its sloping-end cars, front and rear, permitted the rise and descent. (Figure 6.11)

To provide comfortable sleep for a traveler who was obliged to use ordinary passenger cars, Herbert Morley Small of Baldwinsville, Massachusetts patented a hammock in 1889. The traveler could hook it over the back of his own seat and over the top of the seat ahead, dangling his legs through a hole in the bottom. (Figure 6.12)

An even earlier aid for sleepers on the railroad was a headrest that Sebastian Lay of La Porte, Indiana protected in 1881. A drowsy passenger could lay his forehead on a cushioned support, and put his arms on another lower down. Both supports were to be attached to a vertical standard resting on the car floor. Mr. Lay thought his rest would be useful to invalids who could not lie down. (Figure 6.13)

Inventors are still imaginative about the railroad. From South Africa has come a recent proposal for trains to run at high speed on blocks of ice instead of wheels. In his 1967 patent, Dragan R. Petrik of Pretoria indicates a preference for support on flanged metal surfaces, although conventional rails could be used. The vehicles moving under jet or rocket thrust are to put down fresh blocks of ice as these melt and wear away.

Fig. 6.12:
Save the cost of a lower berth.
(Patent 400,131)

Fig. 6.13:
Traveler sleeps sitting up.
(Patent 245,639)

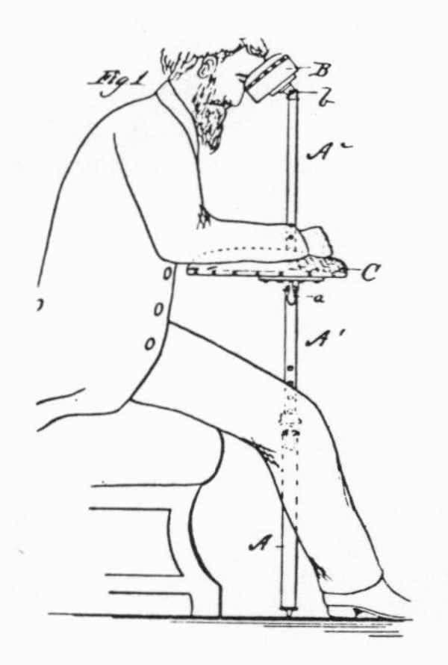

America has never relied solely on steam and gasoline for moving about. There are manpower, dogpower and horsepower. Richard C. Hemmings of New Haven, Connecticut patented in 1869 a velocipede consisting of a large wheel that rolls with the operator perched inside. He rests his weight on a saddle, with his feet free and his head protected by an awning, while he propels the vehicle by turning handcranks linked to traction equipment fitting inside the rim. Mr. Hemmings indicates that numerous experiments proved the ease with which it works and the speed it attains. (Figure 6.14)

Another velocipede, patented the following year by F. H. C. Mey of Buffalo, New York has a carriage seat for the driver and a hollow front wheel turned by a pair of dogs running inside it. (Figure 6.15)

Fig. 6.14:
Crank yourself down the road.
(Patent 92,528)

Fig. 6.15:
Dogpower moves its human freight.
(Patent 190,644)

The "motor for streetcars," devised by Augustus W. Getchell of Cleveland, Ohio is intended to increase the speed and decrease the cost of operation. The vehicle is really a horsecar with the horse inside, facing forward and walking on an endless-belt tread. Mr. Getchell reports in his 1887 patent that the horse does not step faster than its ordinary walking gait, "from which a speed of twelve to eighteen miles per hour is easily obtained." And the animal is protected by roof and walls from heat in summer and cold in winter. (Figure 6.16)

Fig. 6.16:
Horsecar with horse inside.
(Patent 368,825)

As the motorcar arrived, it scared horses that were pulling vehicles along the highways. Henry Hayes of Denver, Colorado proposed that the figure of a horse—with mane and lighted eyes—be attached ahead of a car so that its natural fellows would not be frightened. As described in his 1904 patent, the imitation nag is supported in front by a wheel and pneumatic tire, and its hind legs are fitted into the car's front axle. When its tail is lifted, a storage chamber for fuel, tools and tires is exposed within the rump. A headlamp shines on the roadway from the horse's eyes, and pressure on a bulb sounds a horn inside its head. (Figure 6.17)

Fig. 6.17:
Mechanical horse leads an auto.
(Patent 777,369)

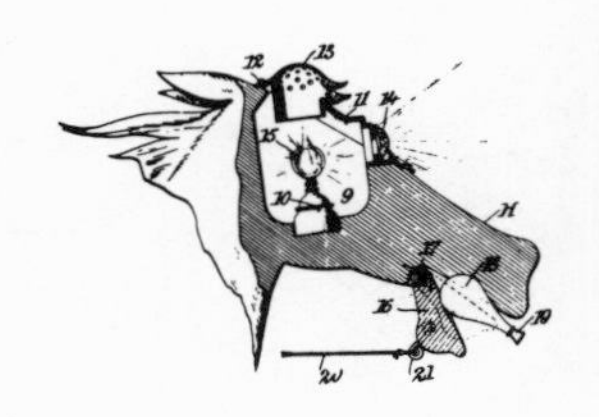

An automobile driven by the force of gravity was disclosed by Isaac Smyth of Chicago in his 1911 patent. As pictured, it has high windows and roof conforming to motorcar style of the period but, unknown to the casual viewer, weights are cranked up inside the vertical sections of the body. When released, these weights propel the car along. During the winding up process (evidently done by hand), the rear wheels may be elevated from the ground. Mr. Smyth confesses that the mechanism may be inadequate for long distances without excessive cranking, but says it is particularly adapted for short city runs. He adds that it might be useful on a theater stage or be copied in principle as a toy. (Figure 6.18)

A tube running from the driver's compartment and ending in a megaphone at the end of the hood was offered in a 1930 patent by Eugene L. Baker of Taunton, Massachusetts as a means of addressing persons ahead, to facilitate the movement of traffic. (Figure 6.19) Heinrich Karl of Jersey City, New Jersey show-

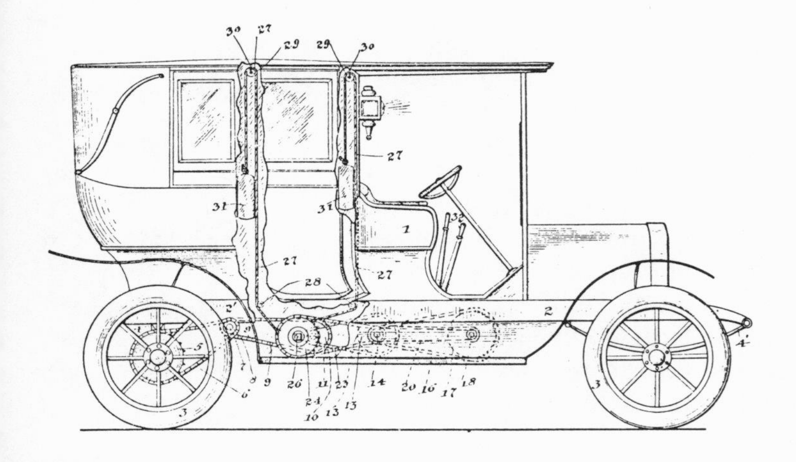

Fig. 6.18:
Driven by force of gravity—inside.
(Patent 995,037)

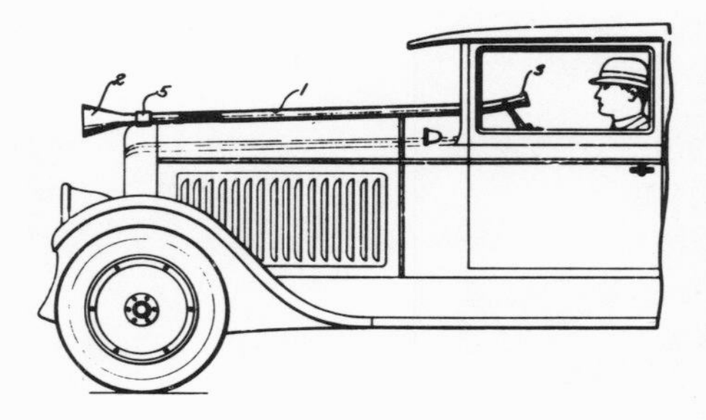

Fig. 6.19:
Tell those pedestrians to move.
(Patent 1,744,727)

Fig. 6.20:
A blanket softens his fall.
(Patent 1,865,014)

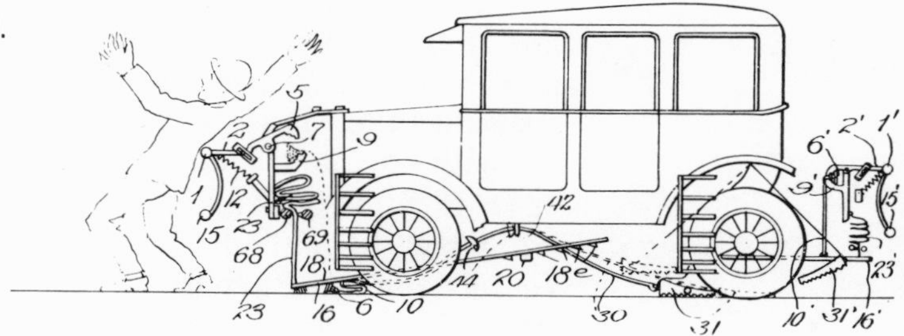

ed great consideration for pedestrians two years later with apparatus installed on the front of a car. It would stretch a blanket, first to prevent a person struck by the automobile from getting hurt by the wheels, and second to soften his fall. (Figure 6.20) Another pedestrian bumper, patented in 1960 by David Gutman, a Philadelphian, was designed to cushion the impact and grasp the bumped person around the hips to keep him from falling to the street. He expected that anyone so struck would be rescued uninjured.

A safety garment for walkers, patented in 1963 by Robert F. Baker of Markleville, Indiana has a green light on the front and a red tail lamp on the back, as warnings for motorists. Under the bulbs are brightly colored patches of material. The front light and patch are at chest height but the level of the rear patch is not indicated. (Figure 6.21) A driver's garment—to keep a motorist cool but modestly covered—is pictured in a 1952 design patent held by Joseph R. Tricarico of Riverton, New Jersey. What might be called a one-armed dickey has collar, a brief shirt front and cloth over the left shoulder, but nothing on the right. From the street, the wearer will look fully shirted. (Figure 6.22)

What was called the "Caster Oil Car" attracted attention in 1953 when a patent for it was issued to Constaninos H. Vlachos and Earl M. Ward. Rumors spread that the castor oil was the fuel, but alas! when *The Wall Street Journal* tracked down Mr. Vlachos in 1957 he explained that the oil was used as a hydraulic fluid and was piped to the wheels; a gasoline engine was required to maintain the pressure. The Castor Oil Car never reached the market and prospective buyers were disappointed.

Fig. 6.21:
Green light ahead, red light in back.
(Patent 3,083,295)

Fig. 6.22:
Full dress for the driver's left shoulder.
(Patent D.167,240)

Water travel has unfortunately been slighted in this chapter. It may be possible to atone for this omission to some extent by concluding with a very old but striking invention—a submarine that can be propelled with one hand and steered with the other. In 1852, L. D. Phillips of Michigan City, Indiana got a patent for what he called "Improvements in Submarine Vessels for the Purpose of Exploring the Bottoms of Harbors, Rivers, Lakes and Seas." The operator, who is shown comfortably seated in the hold, causes the ship to move by turning a crank at the inboard end of the propeller shaft using his right hand. The shaft goes out through the stern by way of a watertight universal joint forming a tiller. With his left hand, the operator can steer the submarine by pushing this tiller up, down or sideways. (Figure 6.23)

Fig. 6.23: Hand-propelled submarine, 1852. (Patent 9,389)

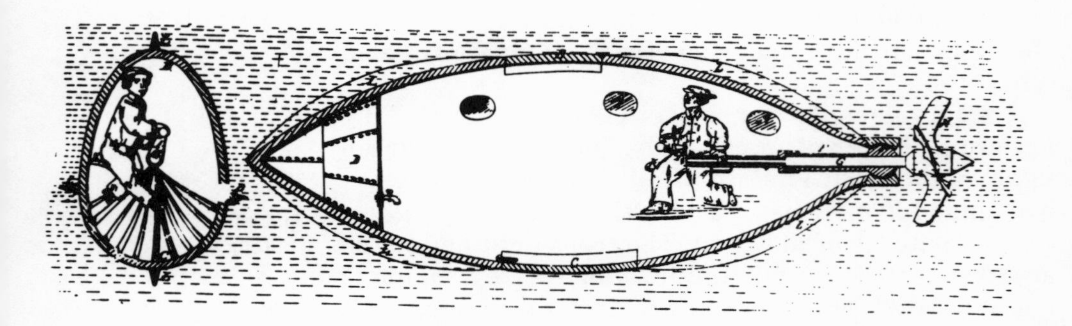

7

Home Is Where the Hearth Is
Ingenuity in the Bathroom

Most Americans hate to get up in the morning and agree with Irving Berlin that it's better to stay in bed. Even bells and rattles are sometimes ineffective, so over the years various alarm systems have been devised that hit sleepers in the face, squirt them with water or drop their feet to the floor.

It was in 1882 that Samuel S. Applegate of Camden, New Jersey patented equipment that would strike a sleeper a slight blow in the face, hard enough to awaken him but not so violent as to cause pain. He

Fig. 7.1:
Get up when the alarm strikes you.
(Patent 256,265)

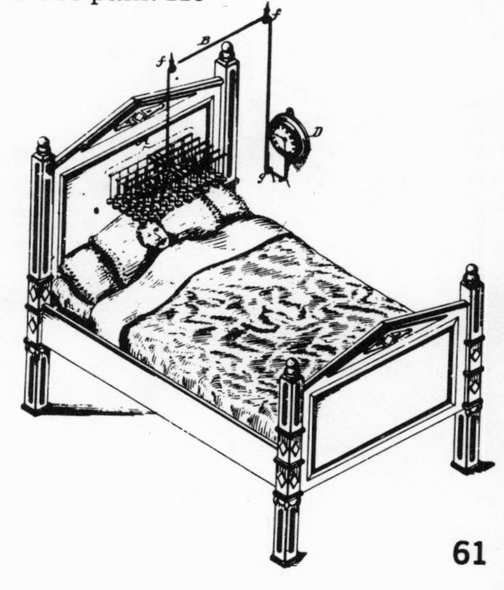

proposed to suspend 60 blocks of cork on cords in a frame. The alarm clock would release them over the pillow so that some at least would fall on forehead, nose, mouth or chin. Mr. Applegate said that connections could also be made with a burglar alarm or to ignite a gas heater in the bedroom at getting-up time. (Figure 7.1)

An apparatus invented by John D. Humphrey of Waterbury, Connecticut for silently striking a sleeper required more accuracy in adjustment. His 1919 patent shows a light rod attached to a table alarm clock with a rubber ball at the end. The rod is adjusted at proper length so the ball will hit the sleeper's head or another selected part of the body and is placed at a nearly vertical angle so that when the alarm goes off it falls on the target. Mr. Humphrey intended his invention for general use but said that because of its silent operation it would be particularly useful for the deaf or for invalids who would be upset by bells or gongs. (Figure 7.2)

Water trickled down the sleeper's neck when George Hogan's alarm went off. At bedtime, the user wound the clock and filled a teacup with water. He then wrapped a perforated hose around his neck and went to sleep. Mr. Hogan, a Chicago inventor, said in

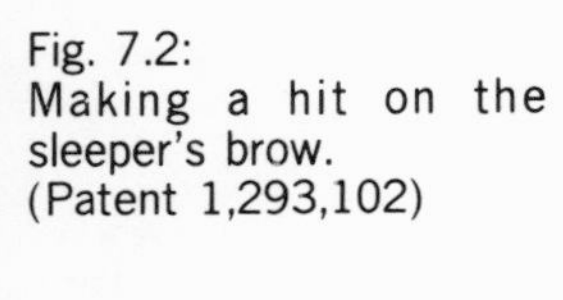

Fig. 7.2:
Making a hit on the sleeper's brow.
(Patent 1,293,102)

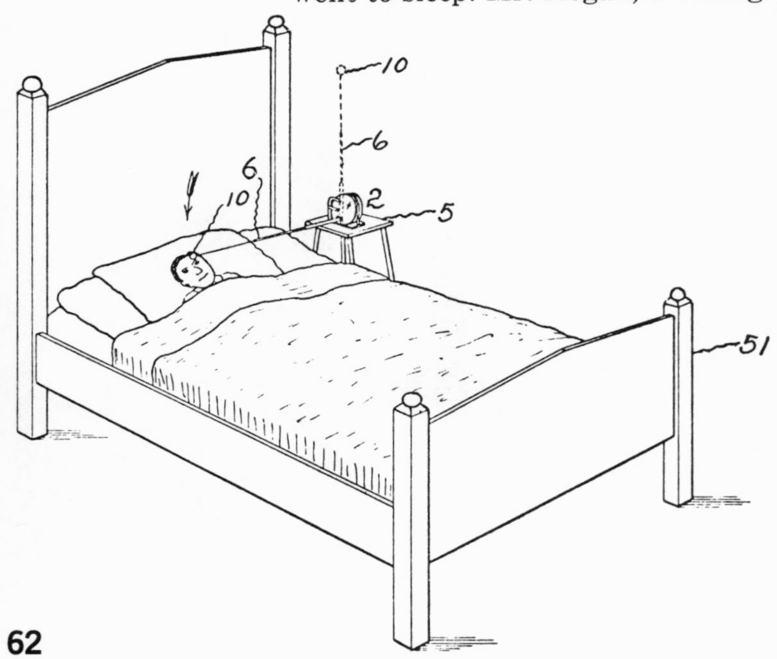

his 1908 patent, "By the use of my invention the person is awakened suddenly and without a shock of any kind whatever."

Ludwig Ederer of Omaha, Nebraska may have been an avid gardener. His alarm bed, patented in 1900, was intended for the man in charge of a hot-house. The bed was normally held level but it was connected to the steam piping in such a way that when the pressure fell to a point endangering the life of the plants it was unlocked and the foot end fell to the floor, causing the occupant to slide or roll off. Mr. Ederer pointed out that when the man refired the boiler and the steam pressure rose, he could lock the bed in horizontal position and go back to sleep. (Figure 7.3)

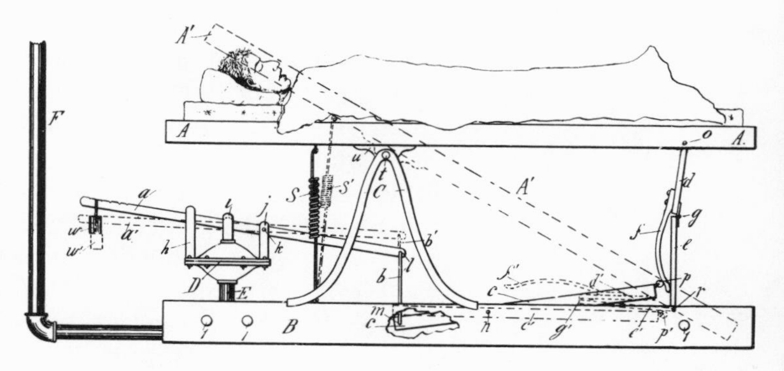

Fig. 7.3:
When steam pressure falls, so does the bed. (Patent 643,789)

For the man who shuts off an alarm and goes back to sleep, discovering later that he has missed four appointments, John E. Coogan of Carbondale, Pennsylvania offered a solution in his 1953 patent. His alarm keeps ringing for an hour—aiming an electric eye across the bed at the sleeper. Only when the subject gets up and the beam is no longer interrupted does the clanging stop. If he lies down again, the noise resumes.

Two inventors offered bed-making aids to the housewife in 1960 patents. Richard W. Nowels of Anaheim, California revealed an electric machine for installation in the box springs, designed to pull the covers tight and smooth. Ernst Wild of Zurich, Switzerland proposed a motorless method. His U-shaped frame, hinged to the bed, could have the lower sheet, top sheet and blanket clamped to it in such a way that when it was pushed down around the mattress everything was neatly in place.

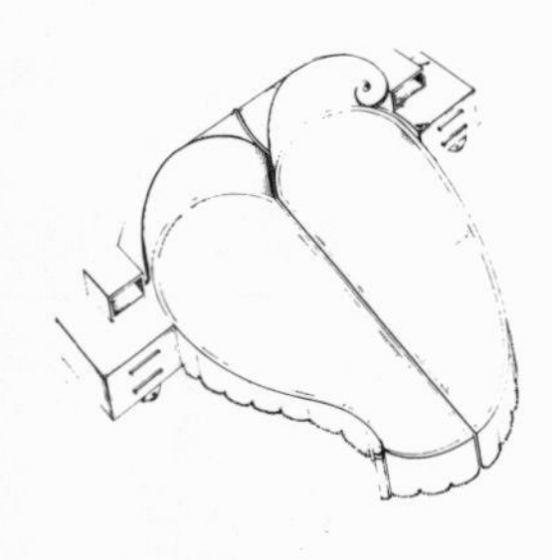

Fig. 7.4:
Heart-shaped bed for bridal couple.
(Patent D.190,989)

A bed's appearance can sometimes be as important as its mechanical structure. J. Vincent Fonelli, a Washington, D.C. artist, got a 1961 design patent for a heart-shaped double bed, intended primarily for brides and grooms. (Figure 7.4) A bedspread which Sidney Shapiro of New York and two co-inventors design-patented 6 years later, simulates a motorboat under way. At the headboard are shown a steering wheel and instruments. The top of the bed serves as the deck of the boat and at its foot is the bow covered with flying spray. (Figure 7.5)

A pillow made with three parallel bars instead of a cushion sounds torturous but it allows a person with a boil on the back of the neck to sleep comfortably. According to the 1966 patent awarded Thomas T. Maru of Honolulu, one bar fits at the nape of the sleeper's neck. The pillow relieves tension and fatigue, lessens the chance of heat rash and protects a woman's hair-do. (Figure 7.6)

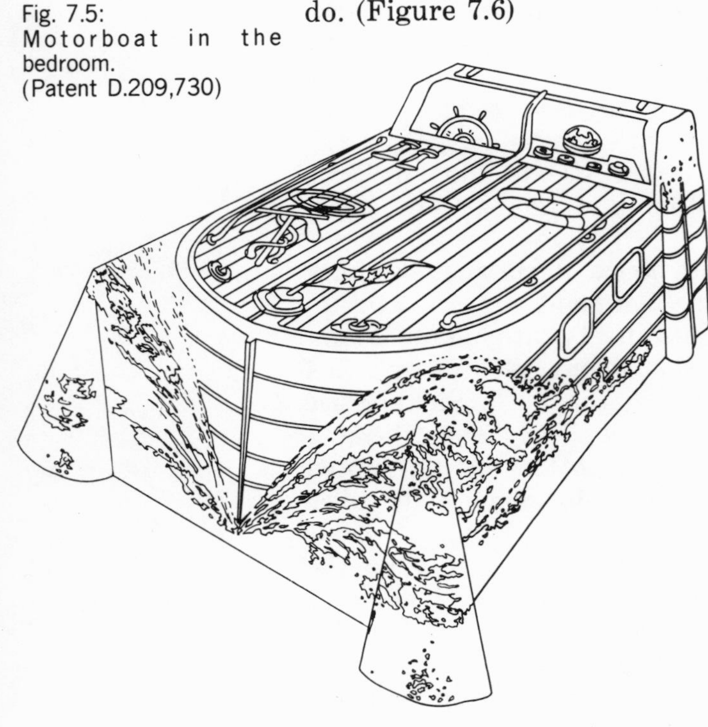

Fig. 7.5:
Motorboat in the bedroom.
(Patent D.209,730)

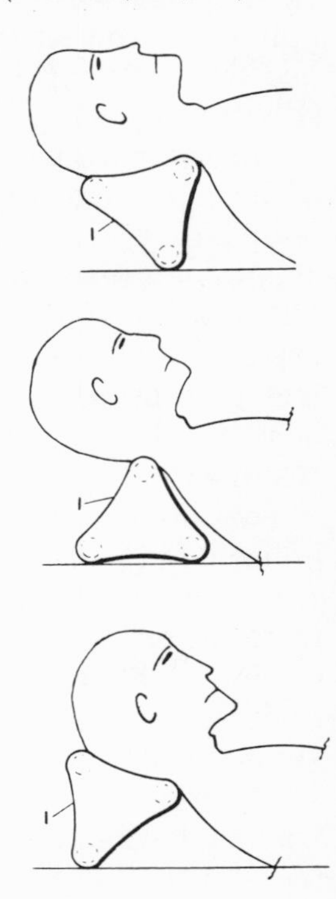

Fig. 7.6:
Sleep with a boil on the back of the neck.
(Patent 3,258,790)

For protection from bedbugs, Frank M. Archer of New York patented in 1898 an electric exterminator that he said would either kill or startle them and cause them to leave the bedstead. Ring contacts were to be installed at joints and other places most inhabited by the bugs so that when the insects moved about they would close the circuit through their own bodies. If the rings were placed on the leg of a bedstead, Mr. Archer said, a bug climbing up would—when it received the shock—more than likely change its mind and return in the direction whence it came.

A simple, mechanical insect trap to be placed under the pillow and operated by body pressure was proposed by Joseph P. Dillin of Ardmore, Pennsylvania in his 1922 patent. It consisted of two plates that were baited by being smeared with raw meat and would crush the bugs caught between them.

An overnight guest—venturing into the bathroom—is sometimes puzzled as to how to operate the washbasin's unfamiliar knobs and buttons. There will be no problem if the householder has installed a recently invented lavatory in which the water starts running when somebody walks up to it. Once the user leaves, the water stops.

The Proximatic lavatory and the antennas that sense a person's presence were patented for American

Fig. 7.7:
When you stand close, the water starts.
(Patent 3,585,653)

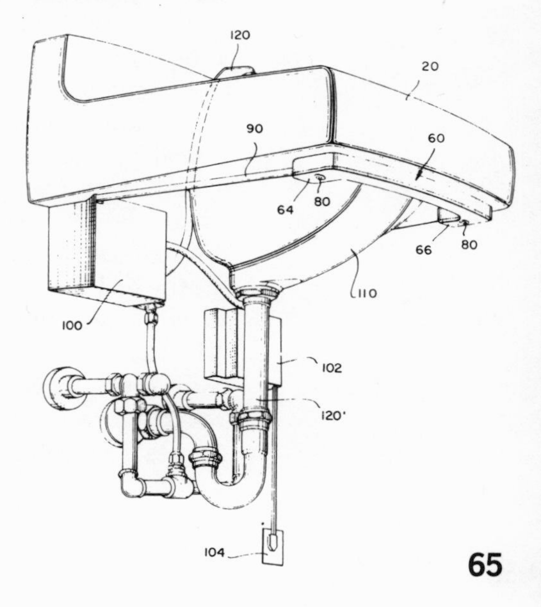

Standard, Inc. in 1971 by Norman A. Forbes and James R. Griffin. The company offered hospital scrub sinks to let doctors wash their hands without touching anything and automatic urinals for public buildings. There was speculation that the system would be extended to the family toilet. Conceivably it could also cause a tub to fill at the approach of a bather. (Figure 7.7)

Even before the privy moved indoors, inventors worked hard on improvements. Giles W. Bower of Enfield, Connecticut and two associates were concerned at the time wasted by workers in earth closets, water closets and dry closets for social conversation, crocheting and other forms of relaxation and amusement. An 1890 patent provided a scale that caused the closet door to remain open and prevented occupancy when weight on the floor exceeded a predetermined maximum—that of one or two persons. The next year Mr. Bower patented a timer that told a passerby how long the privy had been in use.

An electric chamber was offered by Elbert Stallworth of Americus, Georgia in a 1929 patent to provide warmth in cold weather, especially at night after retirement, for the comfortable use of children or others sensitive to the effects of sitting on cold seats. Around the upper edge of the bowl were metallic bands enclosing resistance wires between strips of mica. The seat itself was made of rubber and asbestos. A guard kept children from burning themselves.

An imaginative contribution in the W.C. field is the vibrating toilet seat invented by Thomas J. Bayard, operator of a Chicago hair studio for men. As described in his 1966 patent, the seat is in two halves which may be caused to oscillate separately by the electric motor or both may be set to vibrating at high speed. It is well recognized, Mr. Bayard said, that physical stimulation of the buttocks is effective in relieving constipation. A secondary purpose was to impart a relaxing massage. Besides use in hospitals and in homes, he foresaw coin operation by hotels and motels.

After the kitchen washtub was discarded and superseded by the stationary tub, there was a popular

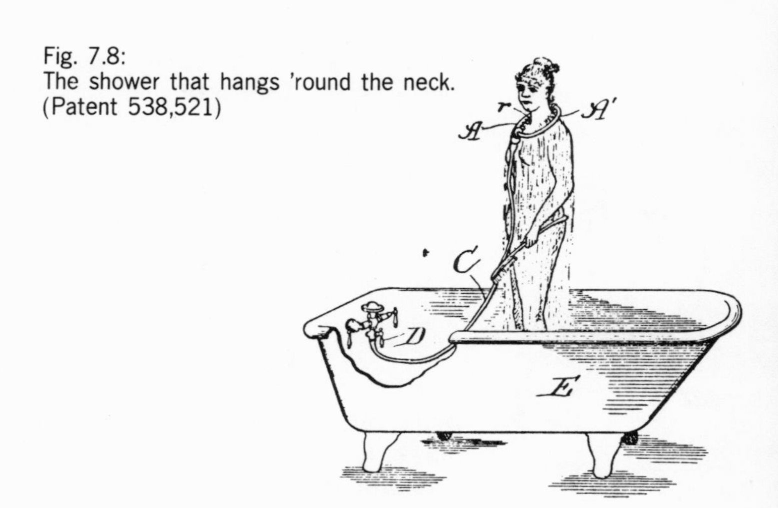

Fig. 7.8:
The shower that hangs 'round the neck.
(Patent 538,521)

demand for indoor showers. In 1895 James Kelly of Chicago patented an apparatus that could be attached to the bathtub faucet and hung around the bather's neck. The part that went around the neck was a perforated metal ring in two sections which could be opened and put in place from one side, without going over the head and wetting the hair. The hinged construction also permitted it to be placed around one leg without being slipped over the foot, and to be quickly removed if the water got too hot or too cold. (Figure 7.8)

In 1913 James Franklin King of Milwaukee disclosed a bathing bag to cover a person from the neck down and to be partly filled with water. The bather could alternately crouch and rise in it, roll on the bed or floor and create miniature sea waves. Mr. King said an attendant could provide manipulation (Figure 7.9)

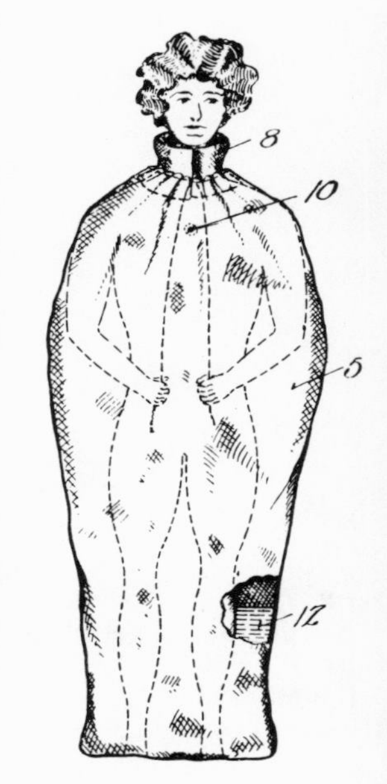

Fig. 7.9:
Take a roll in a bag of water.
(Patent 1,066,121)

The housewife might regard with some skepticism the machinery for scrubbing the bathtub that Robert L. Knight of Coalgate, Oklahoma patented in 1959. Knotted cords threaded through holes in a bucket constitute miniature flails. The bucket is lowered into the tub and detergent is released. Water and the motor are turned on. The bucket moves up and down the tub surface and the flailing cords erase all traces of a ring.

For anybody whose tub runneth over, there is an electric alarm that goes off when a selected high-water mark is reached. Nathan Polikoff of Brooklyn patented in 1952 a box containing a bell and a battery, with an attached float. The box is fastened to the inside of the tub with a suction cup. When water lifts the float it switches on the alarm.

Fig. 7.10:
Kitty stares away the mice.
(Patent 305,102)

For the kitchen and dining room, ingenuity has brought not only many conveniences but protection from pests. To get rid of rats and mice without the use of deadly poisons, John H. Nelson of South Lima, New York conjured up the figure of a cat—drawn on cardboard with several coats of illuminating paint, with phosphorous eyes—so that it would shine in the dark. Besides scaring the pests by its appearance, the cat would exude the perfume of peppermint, which was described as obnoxious to them. Mr. Nelson said in his 1884 patent that his invention was not only an exterminator but a parlor ornament. (Figure 7.10)

Necklaces of bells to be clamped on rats and mice were invoked in 1908 as means to rid the house of them. Joseph Barad and Edward E. Markoff of Providence, Rhode Island said that the tinkling of a bell was as a rule very terrifying to rodents, and that if pursued by such sounds they would immediately vacate their haunts and homes never to return. The inventors baited a trap in such a way that, when a rat poked in its head, a collar carrying bells contracted around its neck. The patent continues:

> The "bell-rat" as it may be termed, then in seeking its burrow or colony announces his coming by the sounds emitted by the bells, thereby frightening the other rats and causing them to flee, thus practically exterminating them in a sure and economical manner. It may be added that the spring-band or collar is not liable to become accidentally lost or slip from the rat's neck because the adjacent hairs soon become interwoven with the convolutions of the spring to more firmly hold it in place. (Figure 7.11)

Fig. 7.11:
Bell the rat and get rid of him.
(Patent 883,611)

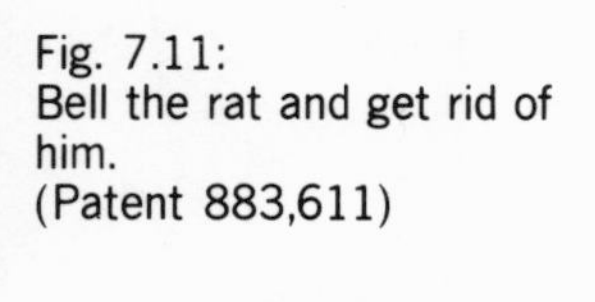

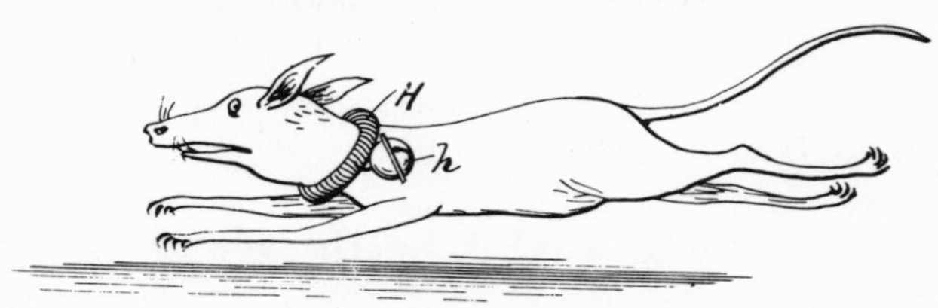

Mice are not only a nuisance in the kitchen but inside the piano as well. Stuart W. Perkins of Falconer, New York revealed in 1967 that the animals get in through slots around the pedals. His solution was a plate with cleats that hold small shutters across the pedal slots. The shutters move up and down with the pedals to which they are attached with rubber bands, and mice are kept out.

An unusual gadget that must have pleased housewives late in the nineteenth century was a combined grocer's package, grater, slicer and mouse and fly trap. As patented in 1897 by Robert Martin Gardiner of Hamilton, Ontario, Canada, it arrived as a grocer's tin can containing such things as coffee and baking powder and was easily converted for other uses. (Figure 7.12) Another patent of that decade—received in 1891 by Albert L. Larkin of Indianapolis—is notable in that the sign it displays offers choice bacon at 7½ cents per pound. (Figure 7.13)

A much more recent kitchen invention promises onions without tears. Wilbert James of New York got a 1967 patent for his Onionmaster, a transparent plastic bag inside which the housewife can peel the herb without being bothered by the fumes. Her hands, an onion and a knife can be inserted through a pair of tubular plastic mittens.

Fig. 7.12:
Grocer's package turns into a trap.
(Patent 586,025)

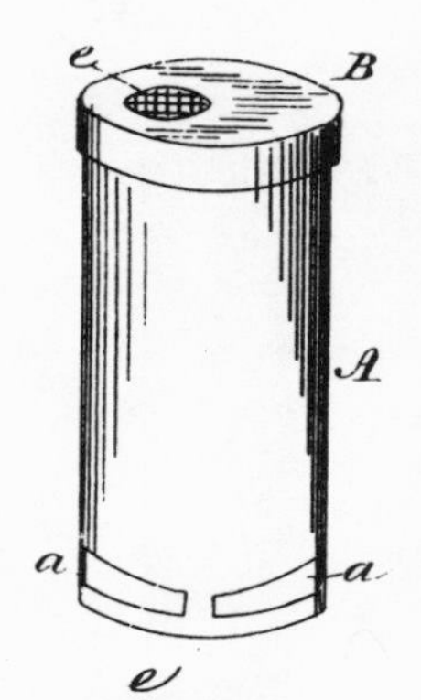

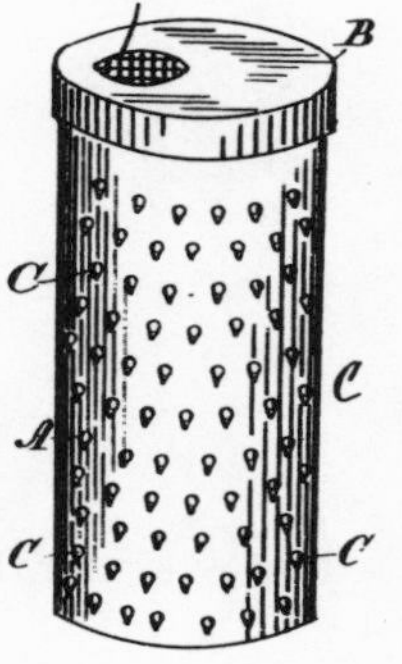

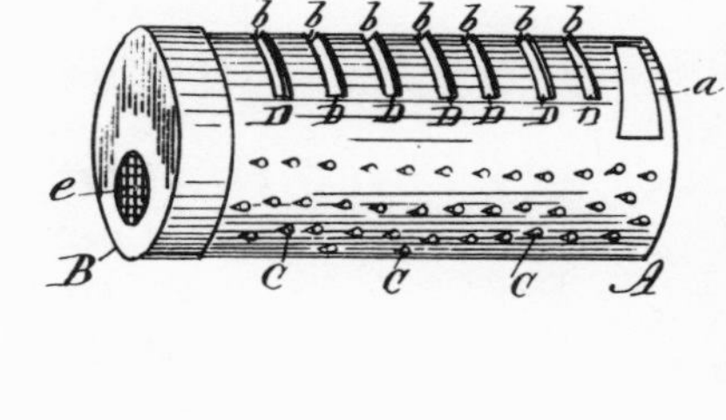

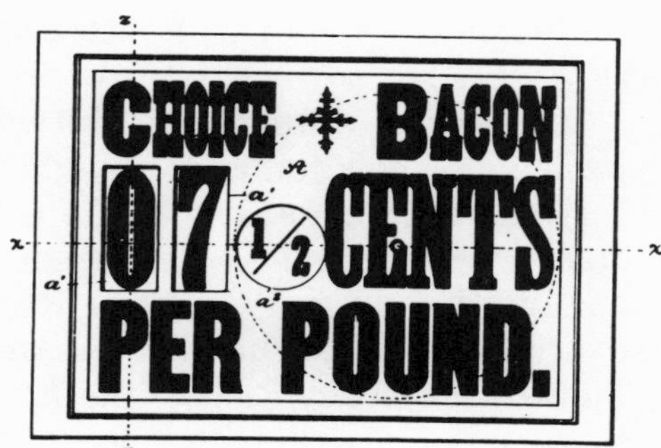

Fig. 7.13:
A sign of the times—in 1891.
(Patent 445,838)

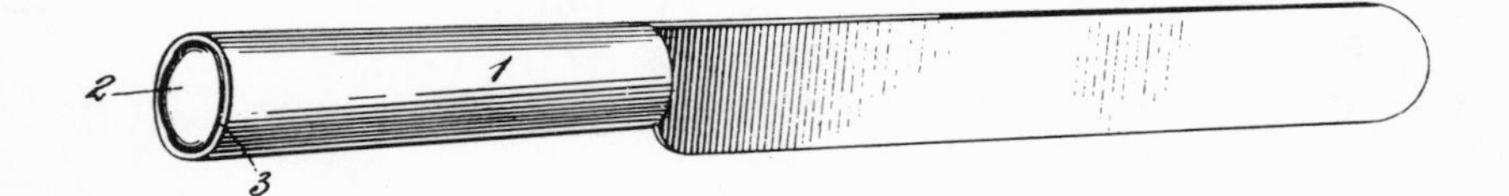

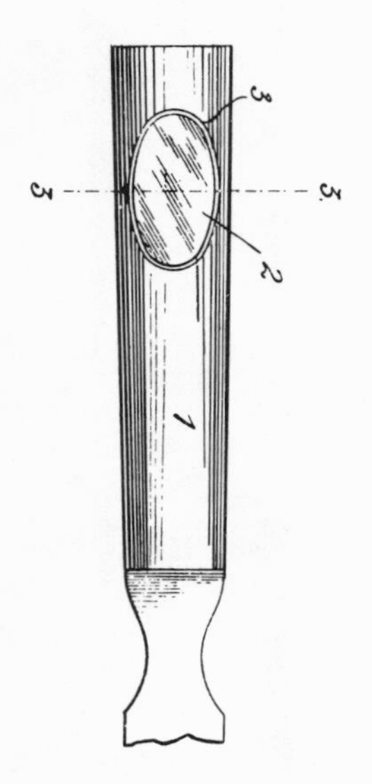

Fig. 7.14:
Inspect your teeth at the table.
(Patent 886,746)

Fig. 7.15:
Wrap the spaghetti with your thumb.
(Patent 2,602,996)

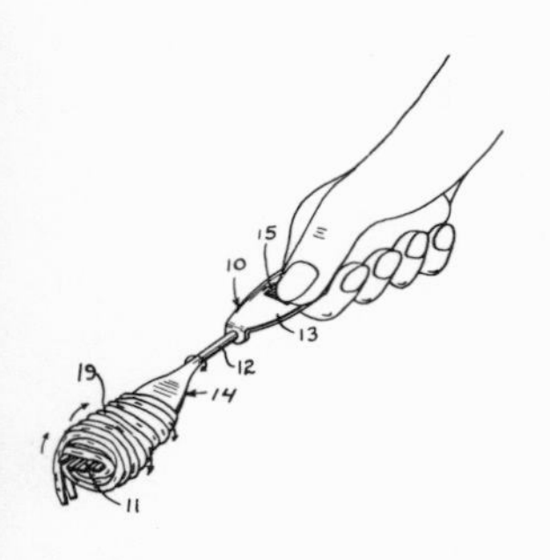

A convenience for the dining room is entitled "Apparatus for Preventing Infants from Sliding Out of Highchairs." John M. Lesh of Santa Ana Heights, California discloses in a 1960 patent that the baby sits astride in a saddle and is held with chin well above the tray. The following year Winifred Willson, a New York writer, patented a combined dining table and dishwasher enabling every diner to clean his or her own plates and silver after every meal or every course by putting them into a compartment near one elbow and switching on the hot water and detergent.

Perhaps to avoid complaints from the children, Clair R. Weaver and Mary A. Weaver of Long Beach, California devised a pie-cutting guide that assures everybody an equal share. Mom can place over the pie a metal pattern with slots for four, five or six equal slices. Then, as shown in the 1963 patent, all she need do is run a knife blade through the channels.

A good many years earlier, people were told how they could inspect their teeth at the table, using mirrors set in their knives and forks. Elmer Walter of Harrisburg, Pennsylvania designed implements with reflectors in their handles. His 1908 patent said often a restaurant patron felt the need of a mirror to discover something lodged in his teeth or to learn whether his lips were clean, but felt too embarrassed to ask the waiter for a mirror. A diner could use a knife or fork without attracting any attention. (Figure 7.14)

An eating implement—intended to settle the problem of how to handle spaghetti—is the revolving fork patented in 1952 by Philippe Piché of Valleyfield, Quebec, Canada. You plunge the fork into a plate of spaghetti and twirl the tines by rubbing your thumb over a wheel in the handle. "As the fork is lifted to the mouth," says Mr. Piché, "the prongs may be rotated to keep the spaghetti properly wound round them." (Figure 7.15)

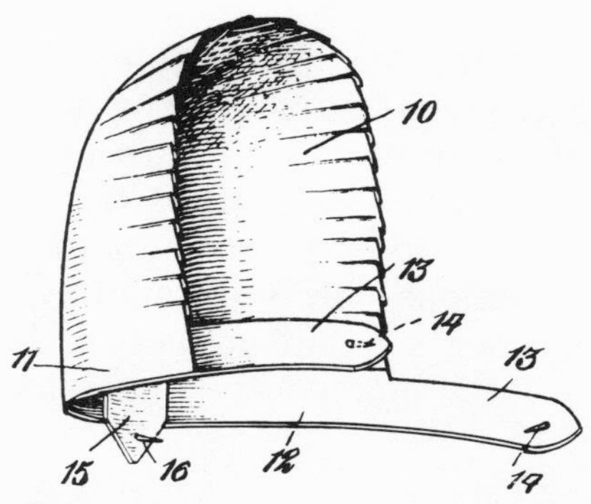

Fig. 7.16:
For protection from grapefruit splash.
(Patent 1,661,036)

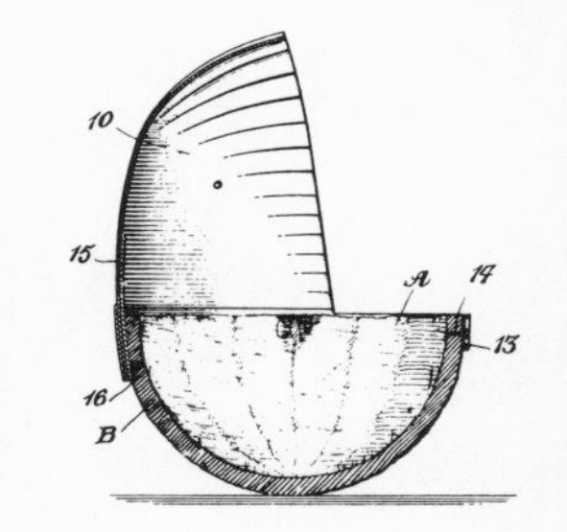

A grapefruit shield was offered by Joseph Fallek of Brooklyn as protection against the mealtime danger of spattered juices. His 1928 patent pictures a waxed paper hood that fits over the open half of a citrus fruit and is held in place by tines. Mr. Fallek said it could be made cheaply enough for one-time use and might well carry an advertising message. (Figure 7.16)

Calling the manipulation of conventional chopsticks extremely difficult for anybody but Orientals, George A. Dawes of Tokyo, Japan patented in 1965 a pair connected by a spring in such a way that their bottom tips always meet. The diner holds one of the joined sticks lightly between his thumb and middle finger and extends his forefinger down the other stick. With this grip he can pick up a single grain of rice from his plate.

A mechanical salad tosser is the contribution of Robert P. McCulloch of Los Angeles, one of the men who bought the London Bridge and arranged to move it to the Arizona desert. According to his 1969 patent, the lettuce, vegetables and dressing are placed in a spherical container with a motor underneath and are rotated simultaneously in two directions, around and around and over and over. The dressing is evenly applied through centrifugal force.

For parties, there is a table decoration that blows up and scatters gifts. As patented in 1971 by Adrien Machet of St. Denis, France, the thing resembles a champagne bottle but holds a guncotton cartridge and fuse. When the wick is lighted, the bomb detonates and small presents are distributed around the table. To protect the table, Mr. Machet recommends that a piece of paper be put under the bomb. (Figure 7.17)

Fig. 7.17:
Exploding table decoration scatters gifts.
(Patent 3,580,173)

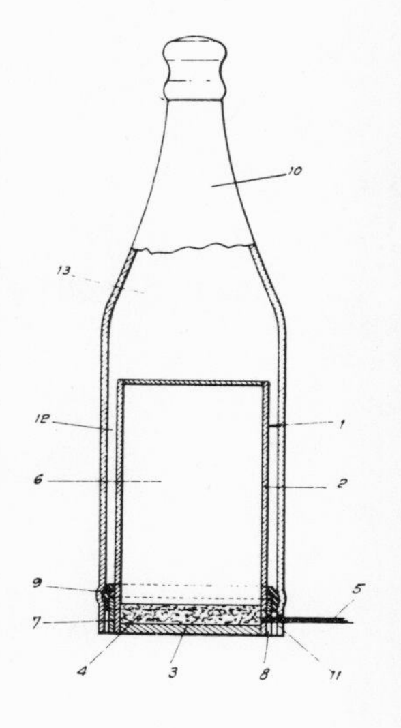

Furniture for the living room can offer unexpected facilities. There is, for example, the double seat through which the hard of hearing can converse. It was patented in 1906 by Virginia M. Hollyday of Baltimore. The two occupants—facing in opposite directions—are separated by what looks like a round, wooden bar but actually is hollow. By putting their mouths against the openings they can talk. (Figure 7.18)

Another piece of furniture can serve as a seat in the living room or as a camel saddle in the desert. William G. Dunn of Clarinda, Iowa points out in his 1964 patent that it has two pairs of legs whose tops cross to form saddle horns fore and aft. A youngster can straddle the seat for indoor play and imagine he is riding across the sands.

Then there are musical chairs such as the two rockers patented in 1952. Roman E. Shvetz of New York, whose music box is wound by a mechanical arm that touches the floor, commented on the "soothing and/or entertainment values" of such furniture. Kenneth E. Haselton of Athol, Massachusetts contributed a detachable music box for children's rockers. He said a youngster could drag it across the floor without catching the plunger in a rug.

Several home devices provide relaxation. Louis Philip McKenzie of Houston, Texas designed a teeter-board on which the occupant, lying supine, can watch television reflected in a mirror overhead. The vibrating couch patented the next year by Forrest Gerald

Fig. 7.18:
Talk to me through the armrest.
(Patent 810,277)

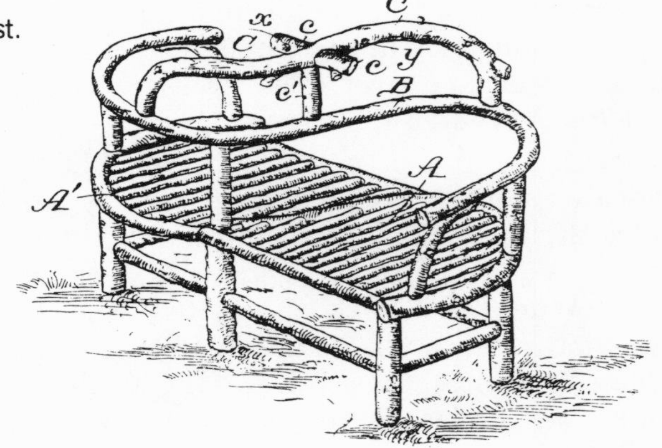

Fig. 7.19:
You can make butter as you swing.
(Patent 383,010)

Johnson of Prosser, Washington gives the occupant a massage and then firmly ejects him. It is intended primarily for coin operation in public places but could no doubt be used in the home. After the allotted time has elapsed, an ejecting bar warns the patron and then the back of the couch pushes him off.

Presumably the swing that operates a churn or washing machine can be used indoors or out. Julius Restein of Philadelphia includes in his 1888 patent a drawing that shows a girl seated comfortably in the swing and looking at a lever as it moves the dasher of the appliance she is helping to operate. (Figure 7.19)

An ingenious piece of indoor furniture patented two years later by Sinclair Arcus of Chicago combines a dining table, a washing machine, a tub and, on top of all, a rocking chair. The occupant of the chair can operate the machine by rocking, and if the chair is removed its water compartment may be used as a bathtub. Mr. Arcus intended his equipment for families with limited income and limited space.

An "artistic creation for interior decoration and human comfort," patented in 1941 by Rafael A. Gonzales of Dayton, Ohio for the Chrysler Corporation is shaped like a palm tree. In summer, brine below the freezing point of water circulates in the trunk forming frost needles on the artificial bark and chilling the room. Mr. Gonzalez says the frost may be made to sparkle by directing colored lights upon it, and air played on the branches may start them waving and cause the leaves to rustle. (Figure 7.20)

Fig. 7.20:
Indoor tree air-conditions the house.
(Patent 2,251,705)

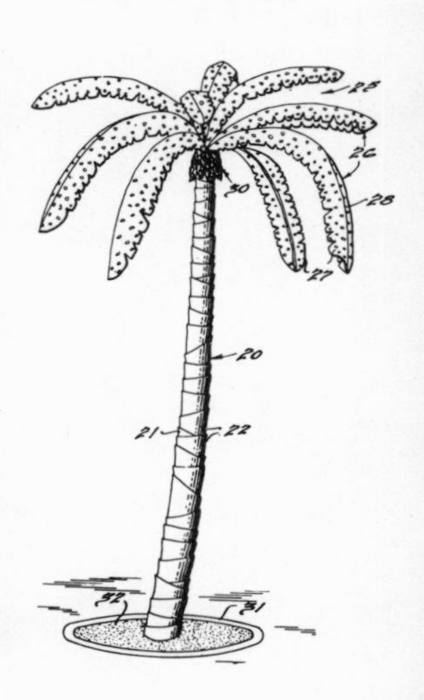

A firecracker fire alarm, its inventor suggests, may be placed in various rooms—perhaps behind drapes. It consists of a tube containing a main fuse and, at spaced intervals, firecrackers attached to it by their individual fuses. Raymond R. Richards of Fall City, Washington says in his 1966 patent that when a fire ignites the main fuse the occupants of the house will receive a series of spaced explosive warnings. As kids would get a bang out of it, he suggests caps over the tube ends, but the main fuse still has to be left protruding through slots in the caps. (Figure 7.21)

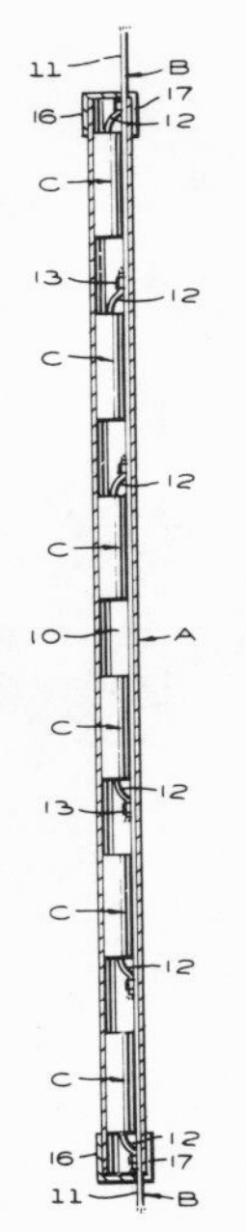

Fig. 7.21:
Fire alarm shoots firecrackers.
(Patent 3,238,874)

A notable invention for use outside the house is the personal fire escape patented in 1879 by Benjamin B. Oppenheimer of Trenton, Tennessee. He said it enabled a person to jump out of the window of a burning building and land on the ground without injury. It consisted of a parachute attached under the chin and a pair of overshoes "having elastic bottom pads of suitable thickness to take up the concussion with the ground." (Figure 7.22)

For the person who wanted to sit on a lawn chair in an area infested by flies or mosquitoes, Thomas M. Prentiss of Boston patented in 1873 a gauze curtain to be suspended from an umbrella with its lower edge resting on the ground. The user could carry the whole thing around by the umbrella handle and when he sat down, place the lower end in a socket or continue to hold it in his hand. (Figure 7.23)

Fig. 7.22:
Personal fire-escape with landing shoes.
(Patent 221,855)

Fig. 7.23:
Safe from flies and mosquitoes.
(Patent 144,792)

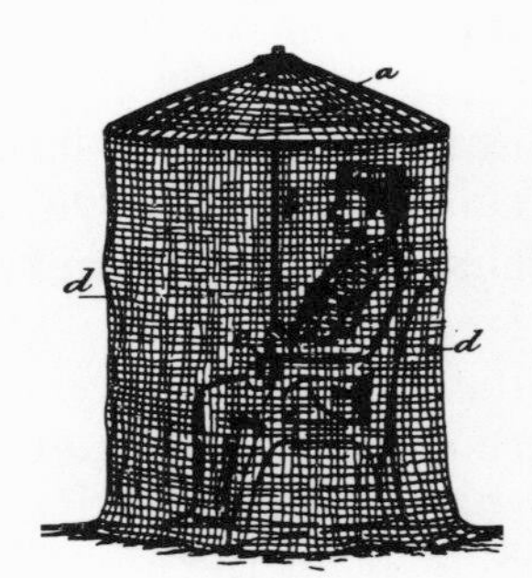

The home gardener, or indeed any digger, was offered in a 1920 patent a shovel equipped to count the strokes. William S. De Camp of Chillicothe, Ohio provided a register to be operated by pressure on a trigger in the shovel handle. The modern householder might use the counter to back up his report to his wife on the labor he has put in. (Figure 7.24)

An automatic house painter, patented in 1971 by John D. Wise of Paterson, New Jersey moves itself along the side of a building, pushing up the outer wall an arm carrying a brush, roller or spray gun. In the same vintage year was Mr. Wise's driverless lawnmower, which he said was able to cut the whole lawn, find its way into the garage and shut off its own motor. (Figure 7.25)

Visitors to the home are fondly remembered, except for the door-to-door salesman. But Frank W. Trabold of New York, in his 1924 patent, offered means by which a commercial caller could become a pleasant memory. The invention was a non-combustible business card of metal, with space for printing the salesman's name and message and with edges that could be folded up to turn the card into an ashtray. On the living room table, it might lead to an invitation to call again. (Figure 7.26)

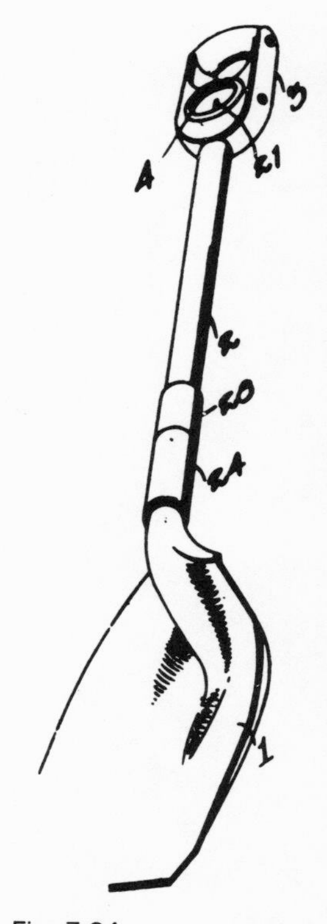

Fig. 7.24:
The shovel that counts its strokes.
(Patent 1,355,541)

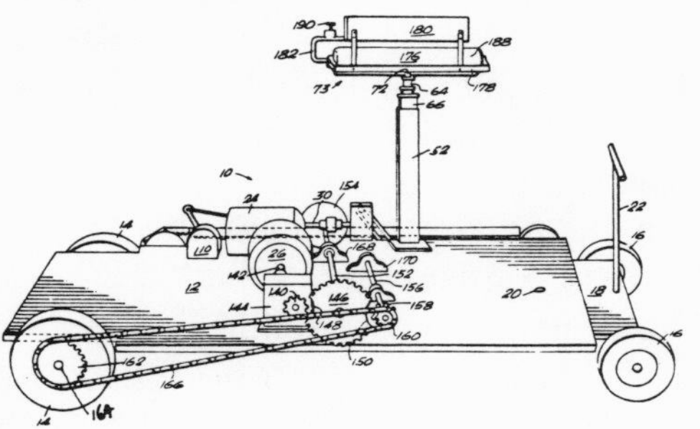

Fig. 7.25:
Automatic house painter runs brush up the wall.
(Patent 3,611,983)

Fig. 7.26:
Visiting card becomes an ashtray.
(Patent 1,508,733)

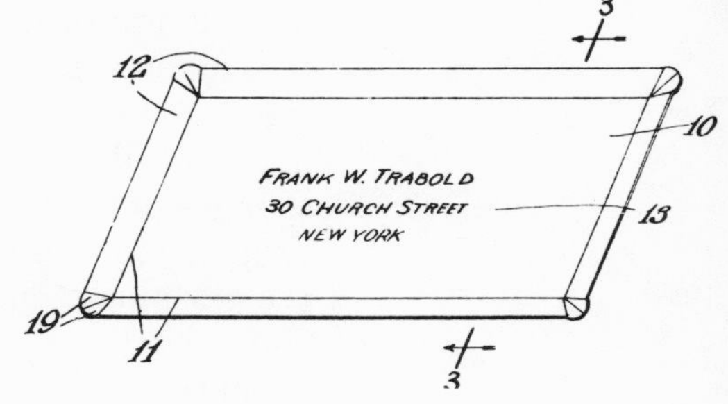

8

For the Sportsman
The Argument at Third Base

Golf is evidently the inventor's favorite sport. An example of sympathetic understanding of the golf addict is the contribution made by Ashley Pond III of Taos, New Mexico—a club to be broken by a player in a fit of temper. The shaft is deliberately constructed to fracture when struck against the ground or a tree "when the anger of the golfer reaches a mercurial height, and wherein the emotion of the golfer requires some physical manifestation to achieve emotional release."

Fig. 8.1:
Club to break in a fit of rage.
(Patent 3,087,728)

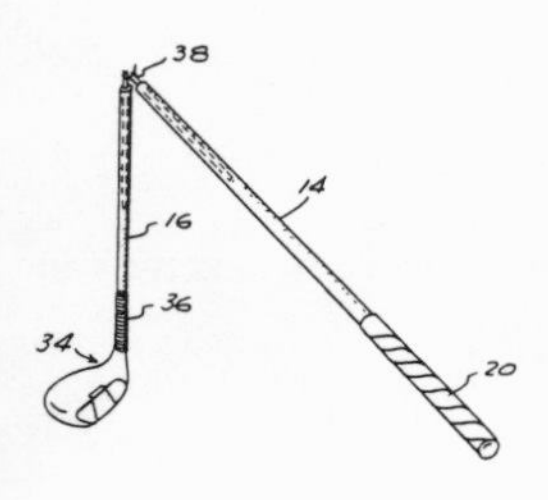

Two sections of the shank are joined by a breakable pin. The 1963 patent explains that after duffing a shot the golfer takes the patented club from his bag, vents his rage upon it and then replaces the broken pin. The regular clubs go unscathed. (Figure 8.1)

A walking golf ball, invented by Donald B. Poynter of Cincinnati, Ohio is reported to have been used by Bob Hope to spoof Jack Nicklaus. A player, after driving onto the green, substitutes "The Incredible Creeping Golf Ball" (its trademark) for the real one and touches its switch with his club, as if he were putting. The walking ball extends its two claws and crawls along and into the cup. (Figure 8.2) By the time the patent came out in 1971, Mr. Poynter had already chalked up sales of 100,000. His company, Poynter Products, Inc., also distributed whisky-

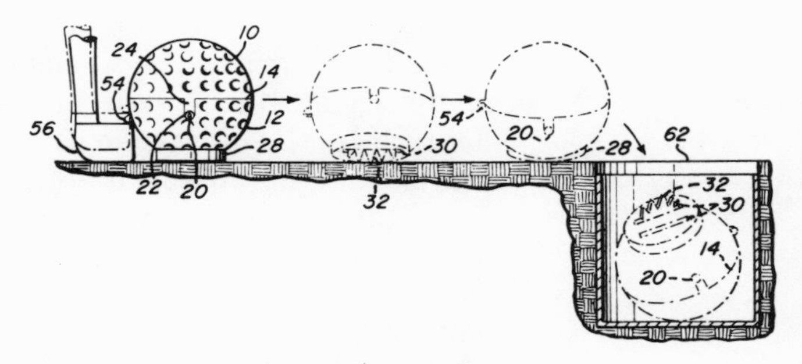

Fig. 8.2:
Ball walks into the cup.
(Patent 3,572,696)

flavored toothpaste (scotch, bourbon or rye) and a soft clock.

Machinery to improve the golfer's stroke takes many forms. Elaborate practice apparatus, patented in 1953 by George M. Troutman Jenks of St. Petersburg, Florida is designed to prevent the golfer from making a false move. His feet are strapped to foot plates, his head is in a cap with chin piece, he wears a belt with three rods linked to the mechanism and the club is gripped by a boom that guides the swing. The machine manipulates the player's hands, wrists, arms, head, shoulders, hips, knees and feet into one integrated motion. He can't turn his head, for example, until after the ball is struck. The mechanism can follow the recorded movements of an expert, who in effect cuts a pattern by which the cams are set. (Figure 8.3)

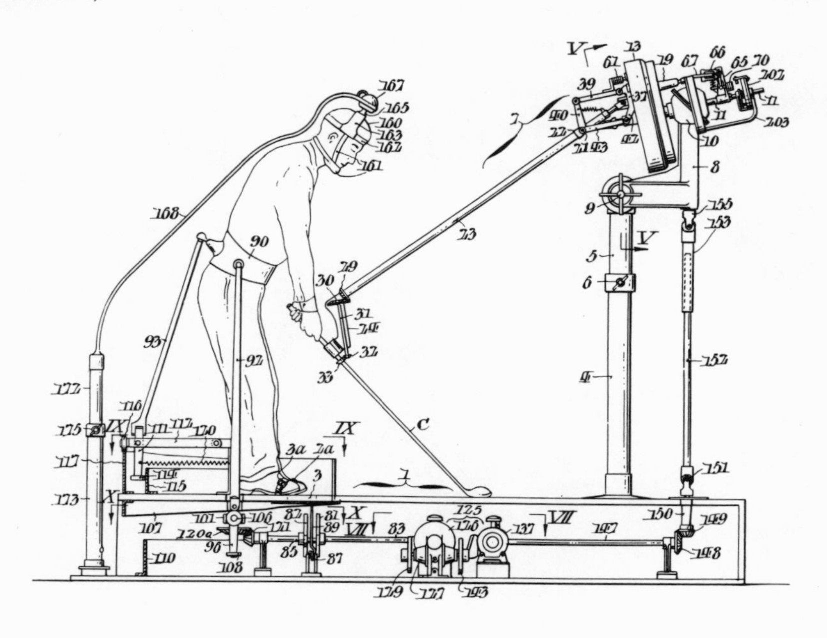

Fig. 8.3:
You can't make a false move.
(Patent 2,626,151)

Once a ball is hit, it often gets lost, and is hard to trace in the grass and weeds. Ellis Miller of London patented in 1918 a golf ball containing radioactive materials. His object, he said, was "to provide a ball which shall be capable of performing with a given expenditure of energy, . . a longer flight than that performed by balls of similar weight and dimensions as at present manufactured." Whatever his purpose, he got protection for a ball that could be traced with a Geiger counter.

Ten years later a golf ball that could be found through the sense of sight, sound or smell was patented by Samuel J. Bens of New York. To create a visible cloud of ammonium chloride vapor, he applied chemicals that would react when the ball came to rest. Another way was to apply a pyrotechnic composition sold for Fourth of July celebrations and called "spit devil." When the ball was struck, there would be an initial detonation followed by crackling explosions when it landed. A coating of elemental phosphorus would emit a glow, even in daylight.

To appeal to the sense of smell, Mr. Bens proposed pleasant odors, such as that from attar of roses and the unpleasant kind, such as the asafetida class. It was also apparent, he said, that dogs could be trained to retrieve balls if they were scented with some essence that would appeal to canine instinct, such as the musks of various animals. (Figure 8.4)

Fig. 8.4:
See, hear or smell a lost ball.
(Patent 1,664,397)

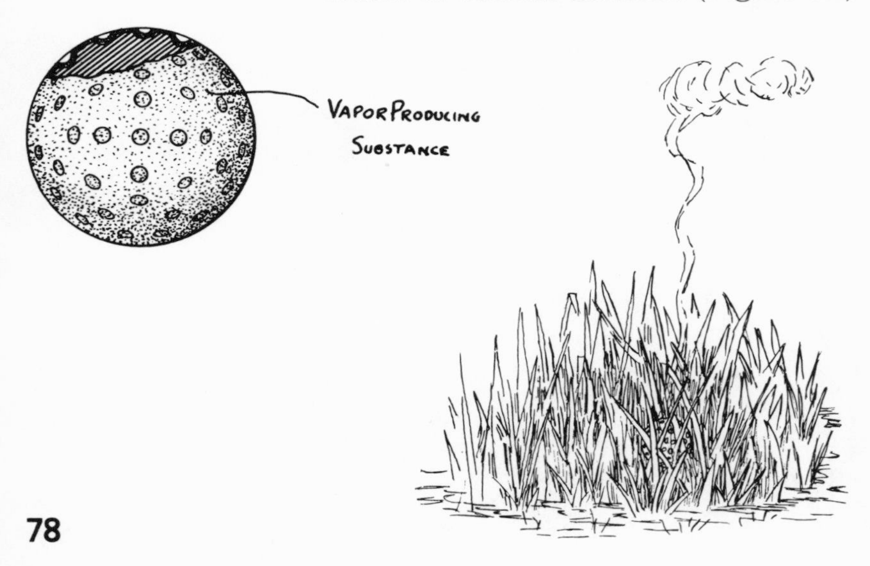

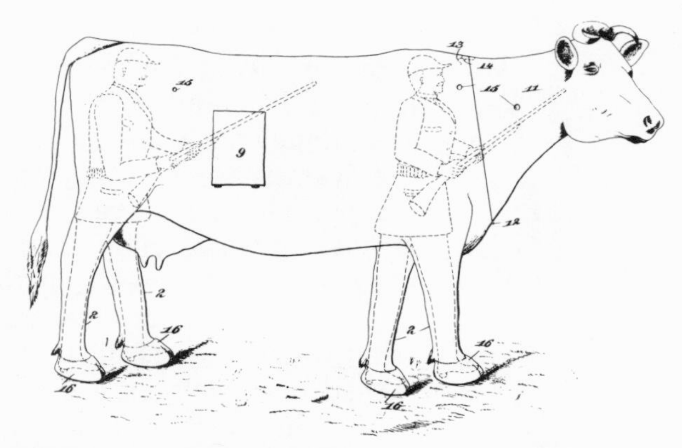

Fig. 8.5:
Two hunters hide inside a cow.
(Patent 586,145)

Hunters must not only trace their game but conceal themselves. In 1897 John Sievers, Jr., of Ames, Nebraska patented a decoy resembling a cow, in which two marksmen could hide—one with his legs in the cow's front legs and the other similarly occupying the hindquarters. Simulated ribs provided a framework and the outer shell was made of hide or canvas. The hunters could walk along inside the cow and open the neck or part of one flank if they wanted to poke out and discharge their fowling-pieces. (Figure 8.5)

A game blind that looks like the stump of a dead tree is described by Harold L. Webb of Milan, Tennessee in his 1961 patent as a comfortable shelter offering a high degree of protection against inclement weather. The outer covering, made of paper pulp or plastic, resembles oak or cypress bark and has simulated roots. The top looks like the cut cross-section of the tree. Inside, the hunter has a floor mat and a stool. He can look out through slots and when he wants to shoot, he stands up, opening the lid. Mr. Webb says his blind may be used not only by duck and deer hunters but by photographers and bird watchers. (Figure 8.6)

Fig. 8.6:
Where the marksman isn't stumped.
(Patent 2,992,503)

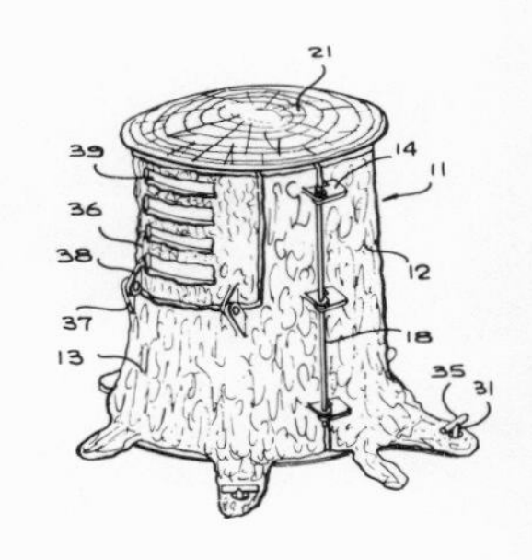

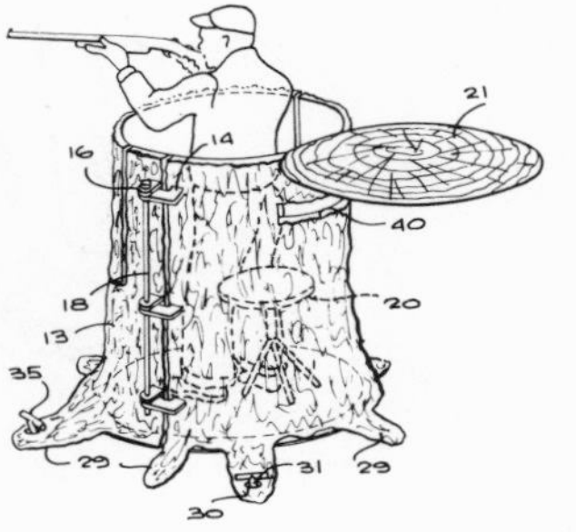

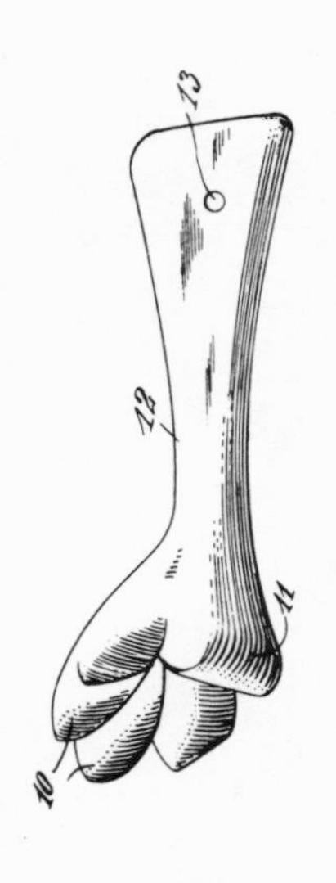

The trapper can press fox or wolf footprints in the snow to encourage such animals to walk into a trap. Raguvald Leland, a Norwegian living in Birch Hills, Saskatchewan, Canada, describes in his 1919 patent the stamps for simulating animal tracks. After the trapper sets his snare—preferably in a trackway the animals have already made—he obliterates his own footprints and presses in the snow the fake animal impressions. There are stamps operated by hand, others at the ends of sticks and a third variety attached to wheels on a sled. (Figure 8.7)

For the outdoorsman, there are patented snakeproof trousers. In 1965 Robert F. Martin of Selma, Alabama disclosed pants with plastic linings below the knee designed to protect the wearer from snakebites. He calculated that a lining five-sixths of an inch thick would keep the fangs from penetrating. For anyone who does not want to bother with trousers, there are snakeproof leggings. (Figure 8.8)

Fig. 8.7:
Making counterfeit animal tracks.
(Patent 1,314,276)

Fig. 8.8:
Pants a snake can't bite through.
(Patent 3,191,185)

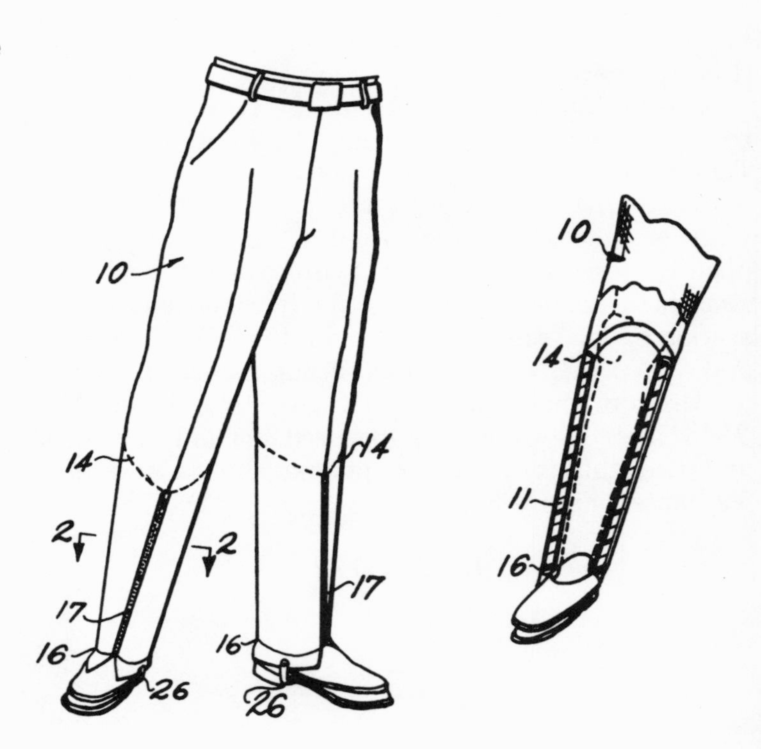

Psychology has been used to bait the angler's hook. William P. Zeigler of Ambridge, Pennsylvania described in 1916 his artificial bait made in the form of a fish with a green head, white and red speckled belly and dark red back. But the main feature is a mirror in the side. Mr. Zeigler explains its effectiveness in these words:

> A male fish seeing his image upon looking therein will appear to see another fish approach it from the opposite side with the intent to seize the bait. This will not only arouse his warlike spirit but also appeal to his greed—and he will seize the bait quickly in order to defeat the approaching rival. In case the fish is suspected of cowardice I may make the mirror of convex form . . . in order that the rival or antagonist may appear to be smaller. In the case of a female fish the attractiveness of a mirror is too well known to need discussion. Thus the bait appeals to the ruling passion of both sexes and renders it very certain and efficient in operation. (Figure 8.9)

Bait that not only looks (to a fish) like an appetizing insect but sounds like one was revealed in 1953 by Stanley A. Wehn of Santa Monica, California. He attached to the fishing rod an electromagnetic vibrator that he said could cause an artificial fly to simulate "the drone or hum of the larger insects or the singing quality of flies or mosquitoes." The vibrator—which can be switched on by finger pressure—transmits the vibrations along the pole and line to the bait. The fisherman can adjust the frequency to reproduce the singing or droning noise of the insect most seductive to the fish he wants to hook.

Fig. 8.9:
Fish sees itself in a mirror.
(Patent 1,180,753)

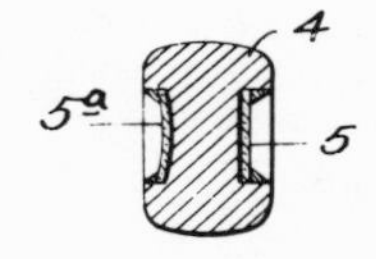

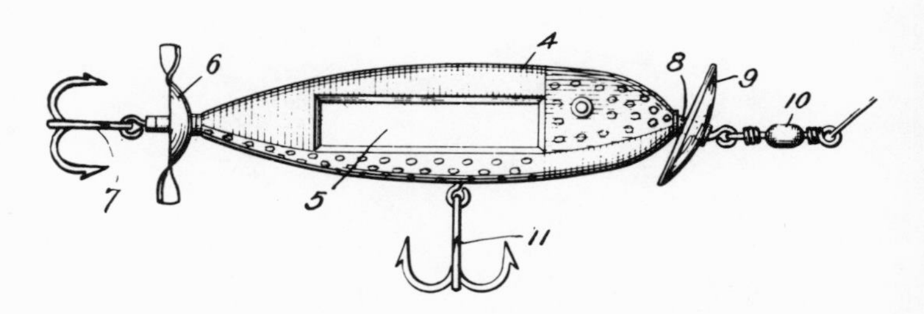

Live bait has received considerable inventive attention. Wilbur C. Fenton of Strawberry Point, Iowa patented in 1952 an electric worm-digger that a fisherman could attach to his car and that would shock the bait to the surface. Two metal rods are pushed into the ground and a neon light shows whether they are properly spaced. For garden use, household current is connected. Away from home, one wire is attached to the car and the other to a spark plug. In the following year James J. Sunday of Detroit patented a combined seat cushion and air pump to reduce the high mortality of minnows in captivity by supplying them with oxygen. The fisherman can install the device in his car or rowboat or use it on a pier. When he sits down, air is pumped into the pail where the fish are. When he rises, a valve opens and the cushion refills. (Figure 8.10)

Man, like fish, can swim under water but usually carries air or oxygen in a tank or has it pumped down to him. Waldemar A. Ayres of Rutherford, New Jersey patented in 1966 artificial gills that were described as enabling humans for the first time to breathe under water. The necessary oxygen was obtained and carbon dioxide was discharged through a membrane permeable to gases but not to liquids.

For walking on water, man has adopted various means—mostly shoes or blades that float and give a grip for forward progress. The buoyant footwear patented in 1971 by Bennie R. Fairchild of Fortuna, California has flexible flippers on the bottom. If the

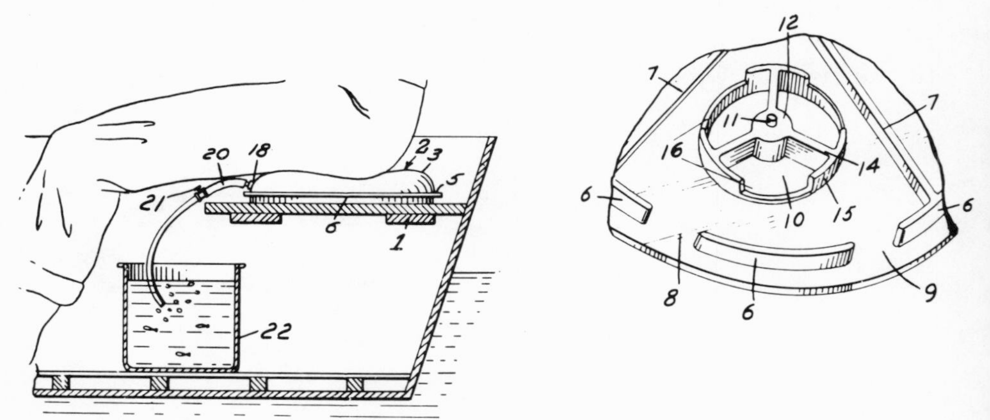

Fig. 8.10:
Pumping air to the minnows.
(Patent 2,664,241)

walker fastens his shoes together, he can rest his weight on a collapsible seat. Marvin W. Brown of Los Angeles in his 1955 patent called his version pontoon water skates and thought they might be useful to Navy and Coast Guard personnel as well as to fishermen and lifeguards. Each skate had blades to be used in pushing forward. The inventor added "hand pontoons" that should be useful when the wearer wants to rest his hands on the water to gain an upright position.

An early patent for a life preserver, issued to Camille Krejci of Scranton, Pennsylvania, includes a drawing of a man in a stovepipe hat, wearing around his neck what looks like a ruff. His body is upright in the water and fish are playing around his feet. The ruff is actually the inflated life preserver, presumably made of rubber and calculated to keep the wearer's head above the water. The man's picture is said to be the likeness of Ulysses Simpson Grant, who was President of the United States in 1870 when Camille got the patent. (Figure 8.11)

Fig. 8.11:
Ulysses S. Grant in a life preserver.
(Patent 100,906)

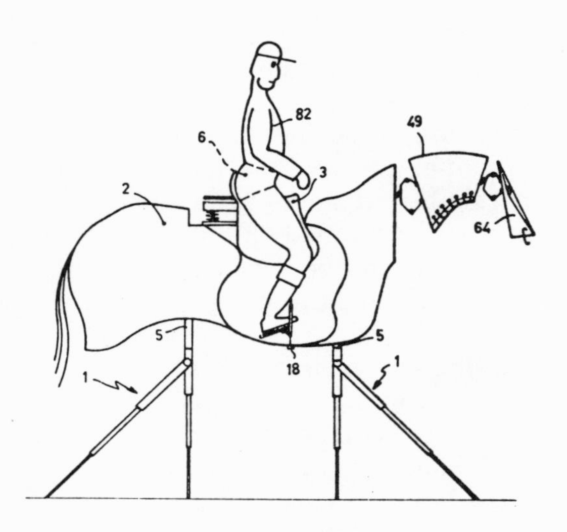

Fig. 8.12:
Learning to ride on a mechanical horse.
(Patent 3,672,075)

On land, sports include horseback riding and horse-race betting—both of which have been mechanized. An "iron horse," patented in 1972 by Matthijs J. Eikelenboom of The Hague, Netherlands instructs a rider without going anywhere. The trunk, neck and head simulate natural movements. The rider sits in a saddle, with his feet in stirrups and reins in his hands, and can practice trotting, galloping, falling off and climbing back on. (Figure 8.12)

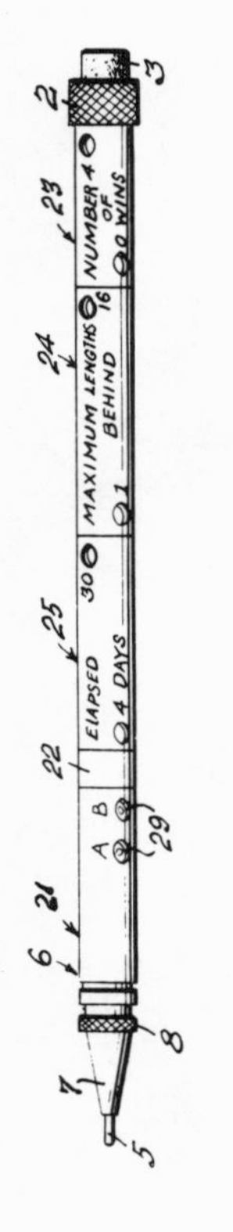

Fig. 8.13:
Pencil computes a horse's record.
(Patent 2,595,153)

Why not computerize the picking of a winner? One device for evaluating horses, patented in 1952 by Arthur E. Strudwick of Long Lake, Minnesota consists of a plate with thirteen buttons small enough to be held in the hand. Each button represents a past performance statistic. With the buttons, a bettor checks the qualifications of each entry and picks the horse with the highest number of credits. The first button is labeled HSR for "highest speed rating." The button marked WC is not what you think but means "won in class." On the same date George V. Malmgren of Chicago protected a forecasting calculator intended to enable a non-professional to weigh and interpret simple horse data. The device is a pencil with several sleeves to be rotated to show such things as how the horse finished in the last race and the maximum number of lengths by which it trailed in any of its last five. If green shows in all of three windows on the pencil barrel, go ahead and bet but be careful if any red is visible. (Figure 8.13)

Ten years later (in 1962) a calculator was patented to forecast—on the basis of past performances—the time a given horse would take to run a given distance. Maxwell H. Hill of Rochester, New York concluded that after the computation was made for each horse entered, the one showing the lowest time was the probable winner.

Of course other animals can race around tracks. Looking for creatures that would not need much land for their contests, Leo F. Buck of Washington, D.C. settled for the razor-back pigs that inhabit the wilds of the South. His 1945 patent describes an oval course around which pigs pursue a tractor pulling an open container of food. The lure is described only as food most appealing to the pigs' natural gluttonous instincts, but at a guess is garbage. (Figure 8.14)

Fig. 8.14: Pigs chase garbage around the course. (Patent 2,376,028)

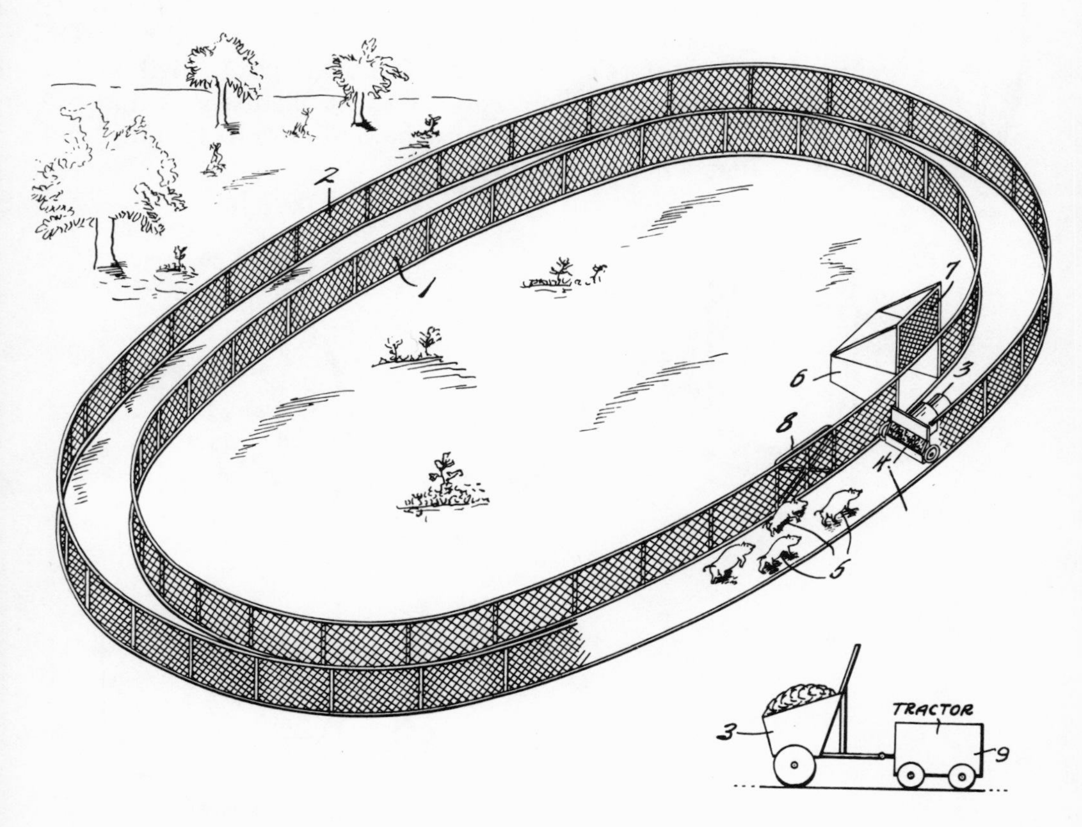

A rodeo performer, James W. Schumacher of Phoenix, Arizona, revealed in 1957 what he called a walking barrel for bullfighting. He said he himself had found it very effective both in fighting and clowning. The inside of the barrel has rope handles so that the wearer can lift it to run or walk. If attacked by an infuriated bull, he can put his feet on a shelf and pull in his head, permitting himself to be butted and rolled about. The recommended material is aluminum, with outside padding made of old auto tires to give the bull protection. (Figure 8.15)

Fig. 8.15:
Barrel for bullfighter to hide in.
(Patent 2,809,035)

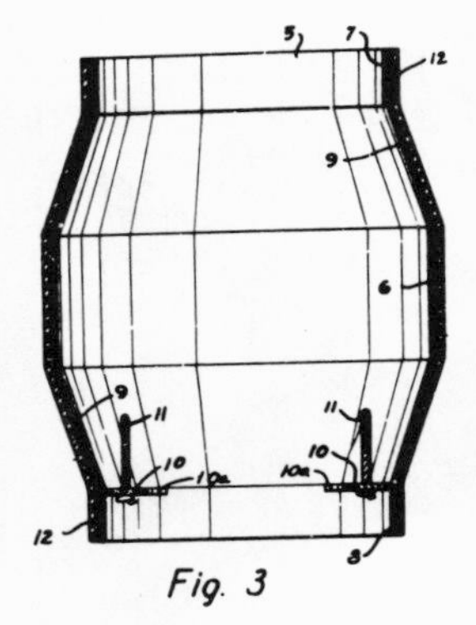

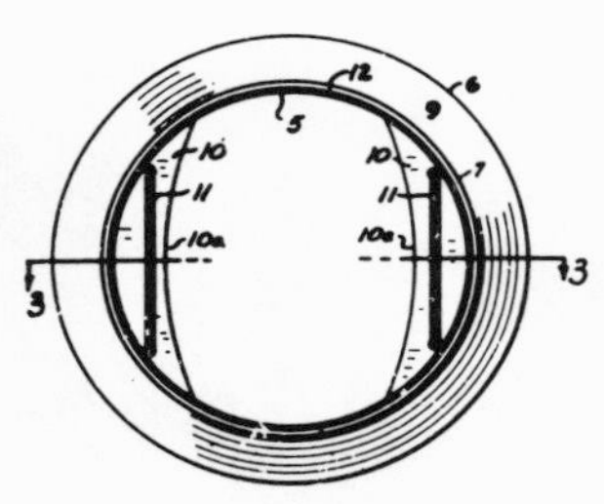

Boxing, another action sport, has received technical aid for many years. In 1895 Joseph Donovan of Chicago, wishing to reduce the roughness and brutality, patented armor equipped not only to soften but to register and ring up all blows landed. He devised a jacket with cushions and numbered electrical contacts over the pit of the stomach, the heart, and over the short ribs on each side. His headgear was similarly padded and wired over the jaws, chin and nose. On the athlete's back, Mr. Donovan put a bell and a register. The bell would ring at each whack and the register would show how many blows had landed and just where. (Figure 8.16)

Fig. 8.16:
Armor scores number of punches and where.
(Patent 543,086)

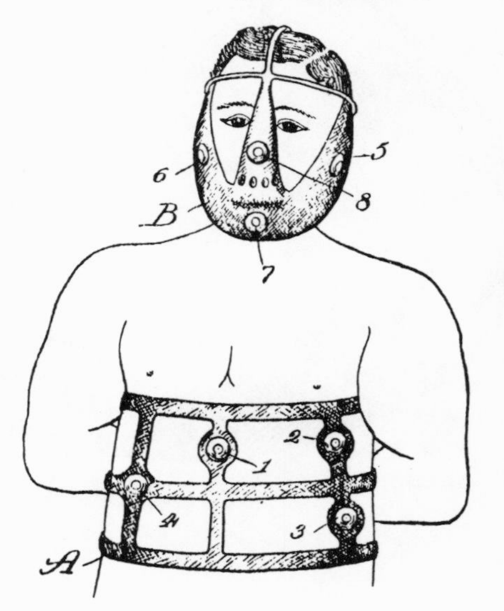

Another method of scoring puts the counter inside the glove. The one patented in 1956 by Willie P. Roberson of Winston-Salem, North Carolina counts the blows a glove strikes, regardless of where. At the end of a match, the judges can read the score inside the wrist band. As the counter can be set to disregard blows under a given force, there is some indication of how effective the whamming has been. (Figure 8.17)

A mechanical sparring partner, patented in 1966 by Jack Preston Nicholson II of Houston, Texas, strikes soft blows because its gloves are filled with rubber. When its electric motor is plugged in, it swings and jabs with its forearms as its shoulders oscillate and its torso bobs and weaves. If a live contestant strikes the mechanical man, the motions of its body and forearms simulate a boxer vigorously returning the punches. (Figure 8.18)

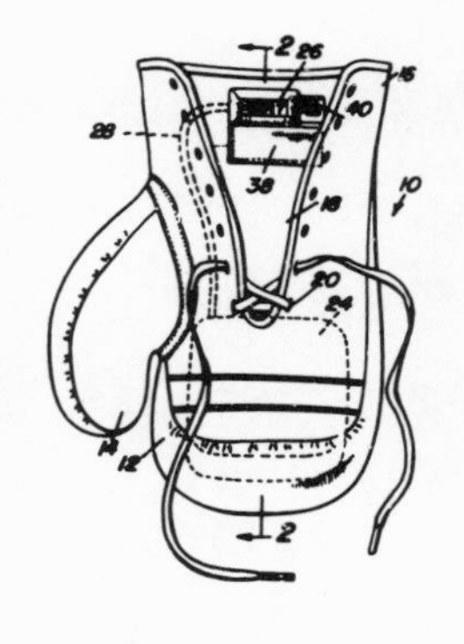

Fig. 8.17:
Glove counts a boxer's blows.
(Patent 2,767,920)

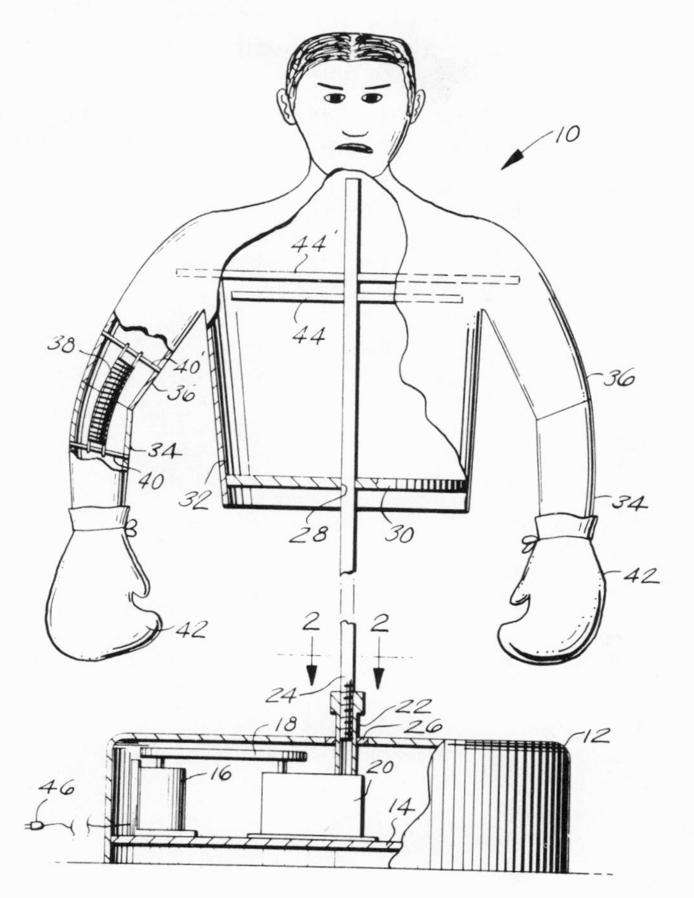

Fig. 8.18:
Your mechanical sparring partner.
(Patent 3,250,533)

Fig. 8.19:
Dummy that teaches karate.
(Patent D.208,057)

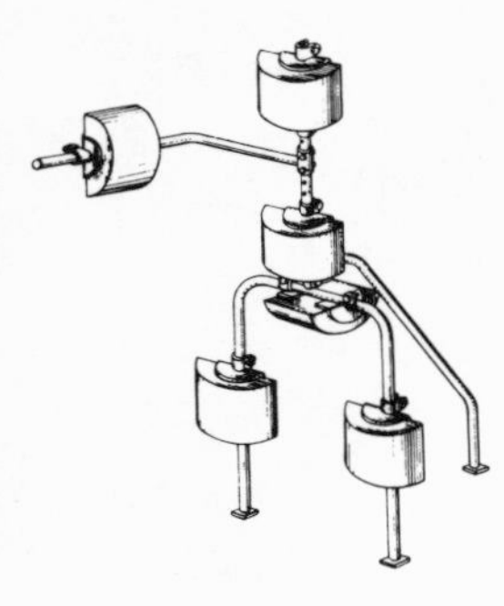

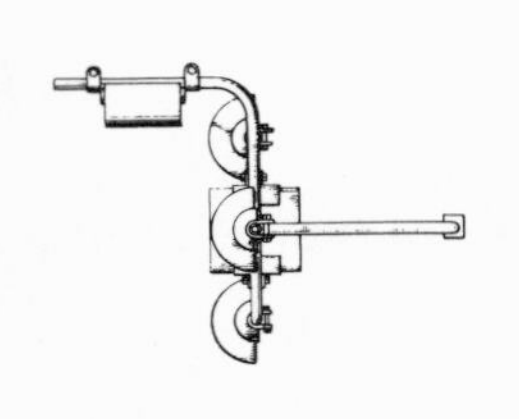

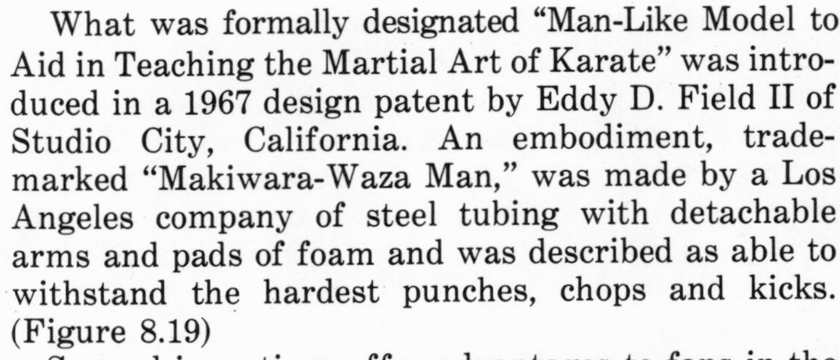

What was formally designated “Man-Like Model to Aid in Teaching the Martial Art of Karate” was introduced in a 1967 design patent by Eddy D. Field II of Studio City, California. An embodiment, trademarked “Makiwara-Waza Man,” was made by a Los Angeles company of steel tubing with detachable arms and pads of foam and was described as able to withstand the hardest punches, chops and kicks. (Figure 8.19)

Several inventions offer advantages to fans in the bleachers. One would let everybody in on the argument at third base. James S. Sellers of Birmingham, Alabama invented an apparatus for transmitting sound, patented by his administratrix in 1962. Microphones are provided for installation under home plate and the other bases so that the remarks of players and umpires in the field will give the spectators a feeling that they are taking part in the game. Besides, the fans can judge a play better if they hear the ball strike a glove. (Figure 8.20)

Clapping mittens are intended to enable spectators at an athletic event to render loud and coordinated

Fig. 8.20:
Home plate is wired for sound.
(Patent 3,045,064)

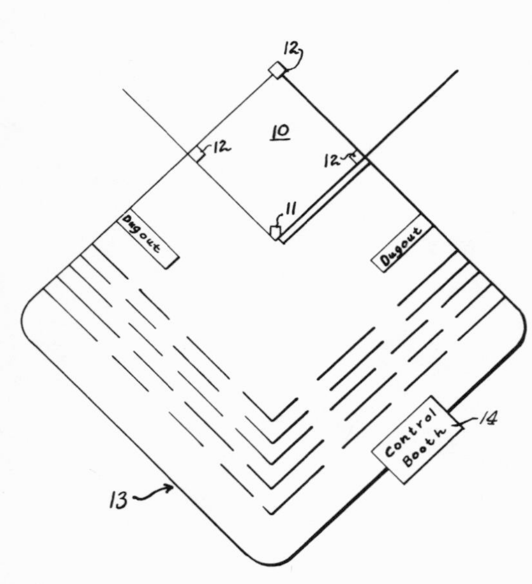

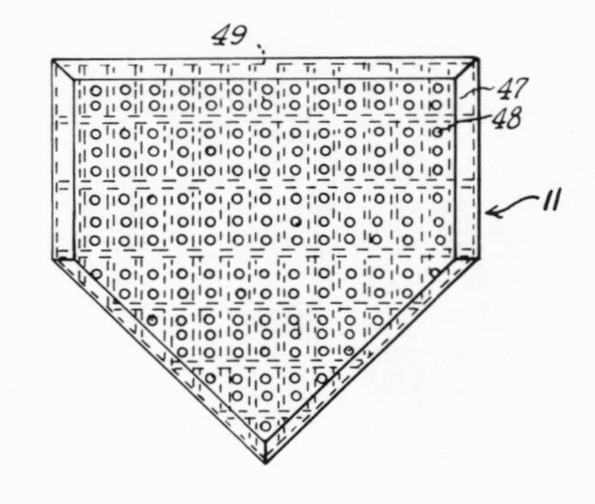

applause for the home team. James M. Crawford, Sr., of Webster, Texas patented them in 1970. The hands fit in resilient pockets at the back. The clappers or blades, which are made of wood, plastic or metal, are regarded as capable of making a lot more noise than human palms.

A West Virginia inventor, Norman G. Foley of Matoaka, has conceived goal posts that will please the fans and not be carried away. According to his 1961 patent, they will shower the crowd with souvenir disks imprinted with the date of the game, the names of the teams and perhaps advertising. When an official releases a spring, the souvenirs will spin around through the air. Presumably that will satisfy fans who might otherwise tear down the posts and use the pieces themselves for souvenirs.

A practitioner of yoga—or a person merely desiring exercise—can stand on his head with an apparatus disclosed in 1960 by Clarence Leonard Horn of Palo Alto, California. There are a sponge rubber pad for the head and a pair of adjustable pads some distance above it for the shoulders. The user grasps a railing and rests his legs against a wall.

One novel means of exercise that should not be overlooked is the mechanical grasshopper with three pairs of spring steel legs that assist a jumper. As explained by George T. and May C. Southgate of New York in 1922, a grasshopper fits around each of the athlete's shoes and is strapped in place. The user (presumably a child) can jump with and land on one or both feet, springing considerable distances, alighting without shock and enjoying the flapping of the wings. (Figure 8.21)

Fig. 8.21: Grasshopper helps you to jump. (Patent 1,402,263)

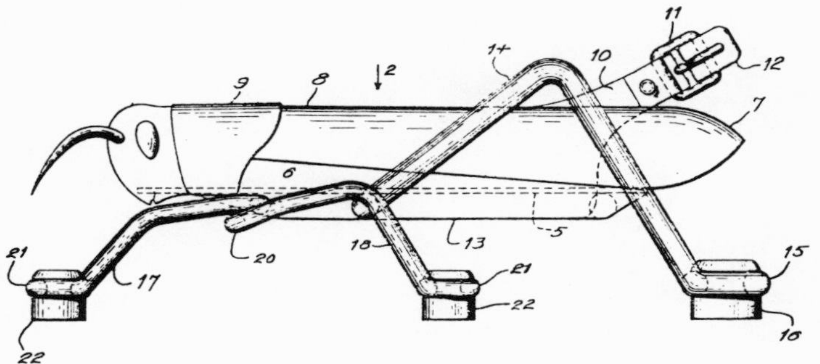

9

Down on the Farm
Artificial Rain and the Mechanical Cow

Agriculture requires a good deal of moisture, and nineteenth-century American technology was called upon to provide it from the sky. In 1880 Daniel Ruggles of Fredericksburg, Virginia patented a method of producing rainfall by exploding torpedoes or other blasting agents in what he termed the cloud-realm.

Fig. 9.1:
A torpedo in the sky.
(Patent 230,067)

Mr. Ruggles gave himself considerable latitude, but one method was to transport the torpedoes by balloon

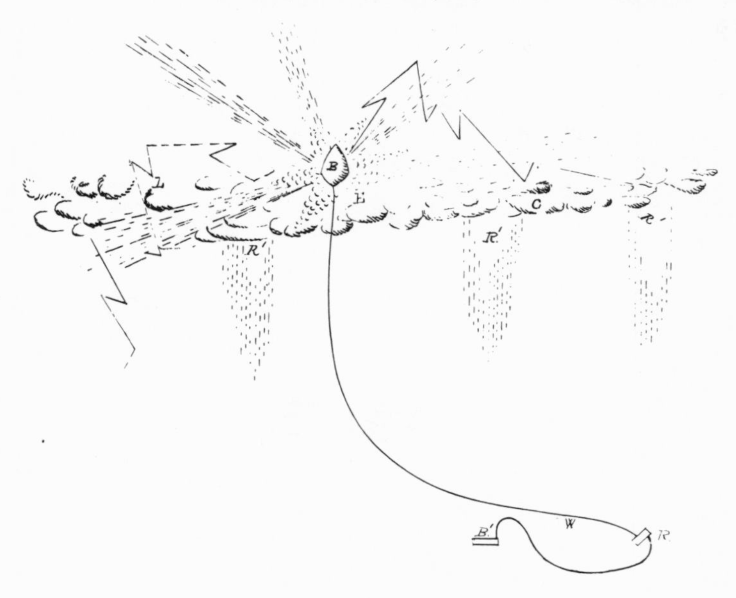

and to fire the dynamite or other explosives in them with electric shocks transmitted by wire from a battery on the ground. He also contemplated having an aeronaut reconnoiter from a regular balloon, trailing torpedoes, throwing them in parachutes and setting them off in various ways.

"My invention," he said, "is based on discoveries in meteorological science, and that electrical force sways and controls the atmospheric realm and governs the movements of the rain clouds, bursting into thunderstorms, dispensing rain and hail, and into cyclones and tornadoes illuminated by magneto-electric forces as prime attributes of matter." (Figure 9.1)

In 1901 William Francis Wright of Lincoln, Nebraska was equally modest in describing his means of controlling the weather—a concussion-mortar. He said it was a

> novel apparatus for firing explosive charges into the atmosphere, for the purpose of producing rain at will and to bring about other beneficial meteorological effects, as the prevention and destruction of hailstorms, tornadoes, drouths [sic], hot winds, frosts, forest and prairie fires and for the modification of atmospheric conditions, largely preventing spontaneous combustion, and for the purpose of sustaining vegetation and for sanitary and other purposes.

Ribs in the barrel of the Wright mortar caused a charge to swirl and turn when it was fired upward into the atmosphere. The mortar could also shoot a time bomb "to explode near the surface or at a low altitude, thus forming upward currents of air and supplying deficiencies of gases necessary to the various effects desired." Mr. Wright said the bomb was not always necessary but a skilled operator could decide when it should be used to regulate the weather. (Figure 9.2)

Poultry attracted the early attention of farm inventors. In 1868 Henry W. Rutt of Reedsburg, Ohio patented a fence that would prevent the ingress or egress of fowls to or from a yard. He took advantage of what he termed a well-known fact, "that hens and domestic fowls never pass over a fence or obstruction

Fig. 9.2:
Bang! and here comes the rain.
(Patent 684,030)

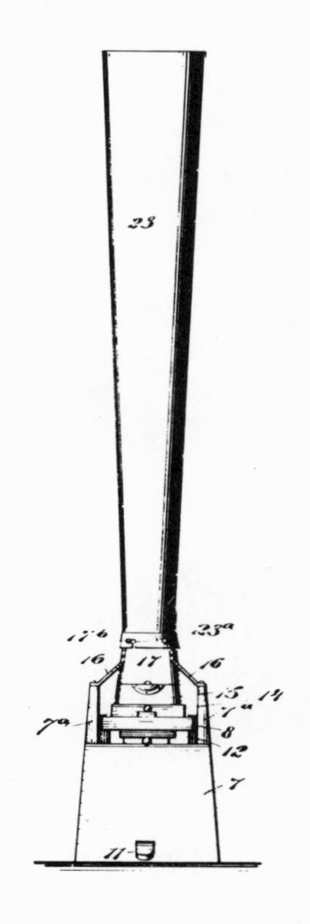

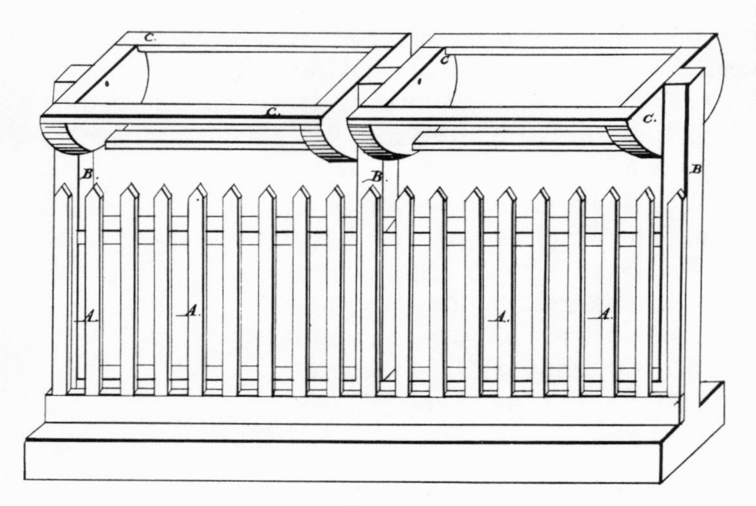

Fig. 9.3:
Fence a chicken can't fly over.
(Patent 77,101)

without first lighting thereon." He built a picket fence in the customary way except that the posts were somewhat higher than customary. Above and parallel to the fence he mounted between the posts long, two-edged perches that could rock back and forth. A hen trying to get over would land on the nearest edge and be thrown back on the same side she started from. Thus a fowl trying either to escape from a pen or to invade a garden would be foiled. (Figure 9.3)

A poultryman likes to know which of his hens are good producers. In 1910 Stanley A. Merkley of Buffalo, New York patented his egg-marker whereby, he said, "the laying capacities or qualities of each hen in a hennery can be easily ascertained." Markers in various colors and combinations were attached to a hen's vent, and as she produced an egg they would imprint it crayon-fashion with a distinctive design. Mr. Merkley specified marking compositions insoluble in water so that the egg would not be smeared. (Figure 9.4)

Fig. 9.4:
Mark the egg as she lays it.
(Patent 970,074)

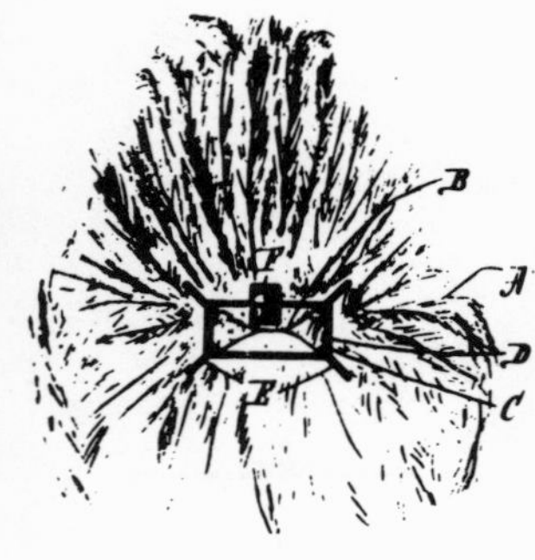

To separate the egg-layers from the non-layers, Nathaniel M. Bain of Oak Grove, Oregon invented his segregating nest. As explained in his 1921 patent, it will trap a hen unless she lays an egg. When she enters the housing and sits on the nest, a door will be locked and will remain so, keeping her confined unless she produces. If she delivers, the egg rolls onto a door-releasing mechanism and she is free to depart. As she emerges, a drop of pigment falling on her body automatically identifies her as a layer.

Some poultrymen probably count their eggs before they are hatched but most of them like an actual production tally. John Hazeltine of Nelson, England in a 1954 patent harnessed the hen so she would make a record of her own egg-laying performance or at least of her visits to the nesting-box. A saddle attached to her wings carried overhead a revolving rubber stamp. Each time she left the nesting-box, the stamp moved over an ink pad and printed her number on a strip of paper. To compensate for the bird's rolling gait, the framework guided the stamp to the proper spot. And Mr. Hazeltine arranged to have each egg that rolled from the box cause a confirmatory puncture in the same recording strip.

A decade later a Swedish inventor disclosed an actual egg-counter to be strapped on a hen's rear. Hans Eugen Birch-Iensen of Billeberga called it better than using special cages or nests to record production by trapping the eggs, as it allowed the birds to live in flocks and follow their natural habits. His harness suspended a cylindrical counter beneath the hen's tail with a pair of prongs to be moved by each egg as it was laid. (Figure 9.5) Mr. Birch-Iensen was critical of an early egg meter, patented in 1932 by Alonzo H. Greene of Fort Worth, Texas, as he said its wearer—while trimming her tail-feathers—was likely to trip a false alarm and throw off the tally.

Fig. 9.5:
An egg-counter strapped under the tail.
(Patent 3,123,044)

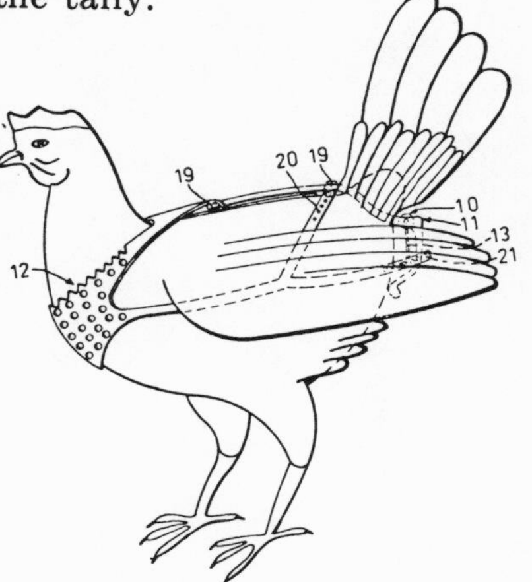

Neatness is desirable in the poultry shed. Hubert C. Hart of Unionville, Connecticut reported in 1915 that rods were commonly used as poultry roosts and that the droppings caused considerable labor when the floors were cleaned at regular intervals. His solution was a long roosting rod with a movable belt directly under it. When the rollers were turned once a day with a hand crank, the belt moved along and was neatly cleaned by a scraper. The floor remained clean.

A mechanical hen was created to dispense eggs in cafes and restaurants. She sat in a nest on the counter and—when a lever in front of her was pressed—she lifted her head, opened her beak, cackled and released an egg at the rear. Gustaf A. Almstrom and Peter Christenson of Northport, Washington, who got a 1907 patent, said that with slight modification the hen could be turned into a vending machine. (Figure 9.6)

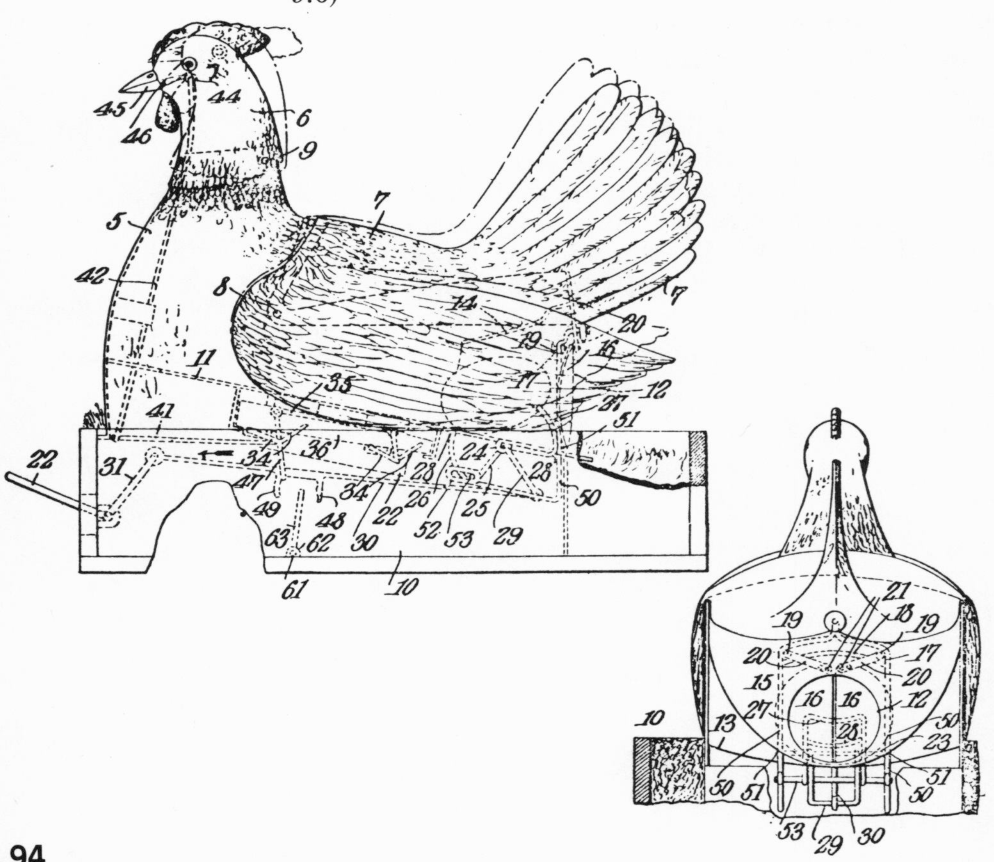

Fig. 9.6: Mechanical hen lays eggs in a restaurant. (Patent 868,632)

That important member of the farm community, the cow, has inspired many technical advances. One recorded in an 1887 patent is aimed particularly at the cow's human companion, the milkmaid. Allen B. Cowan of Hall's Valley, Ohio originated a milking stool to be worn on the milkmaid's back below the waist and to provide instant support whenever she wanted to sit down.

The stool hung down behind the maid, supported by straps around her waist, leaving both her hands free to carry two pails. When she was ready to milk, she leaned slightly forward and it swung beneath her without being touched by either hand. If the cow should move away a few feet or start kicking, the maid could get up and look after the buckets without paying attention to the stool.

The lower end of the board at the back of the stool was swiveled. Mr. Cowan explained the maid could walk into a stall that was too narrow for her to sit down facing the cow. She could settle sidewise and then swing around. (Figure 9.7)

Fig. 9.7:
A milkmaid attached to her stool.
(Patent 359,921)

Masks have been designed for milkers, and others for milkees. To guard the dairyman from being lashed

Fig. 9.8:
A catcher's mask for the milker.
(Patent 1,290,140)

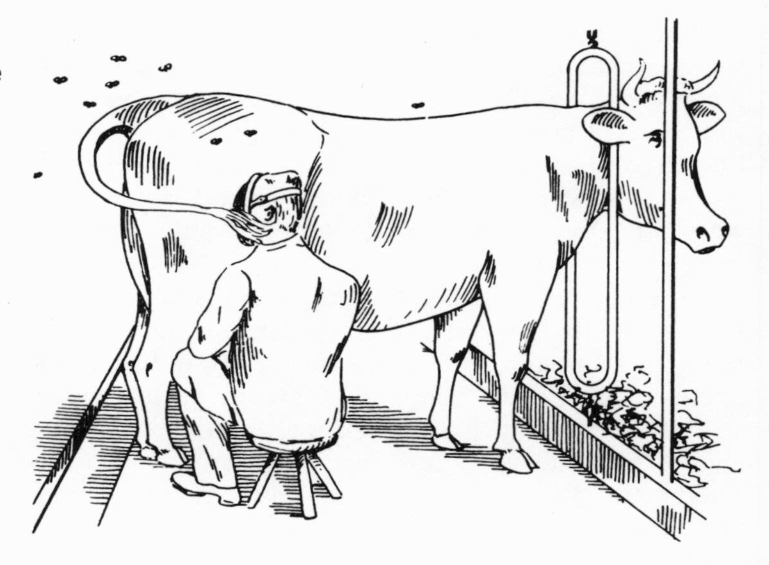

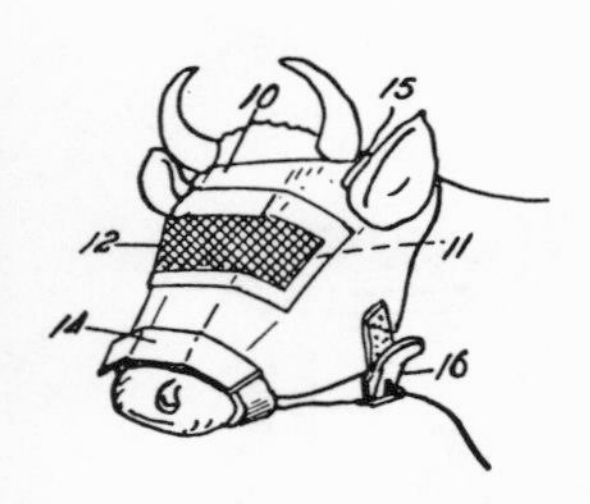

Fig. 9.9:
Keeps those flies out of her eyes.
(Patent 3,104,508)

Fig. 9.10:
Cow's tail is locked out of the way.
(Patent 2,619,936)

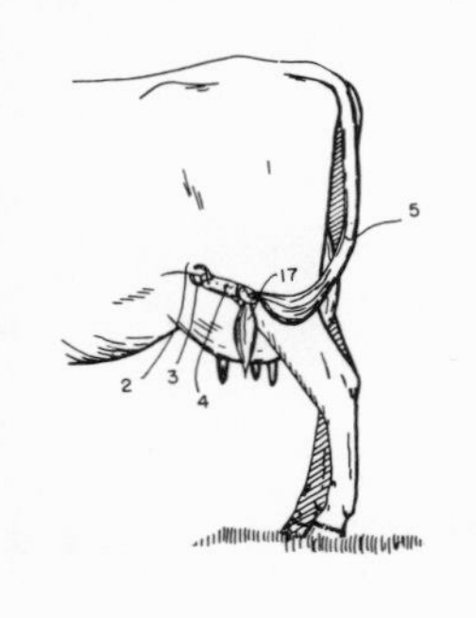

in the face and eyes by the cow's tail during the milking operation, Frederic W. Elleby of Modesto, California invented a guard with much closer wiring than a catcher's mask because—as he explained in 1919—a cow's hairs are much finer than baseballs. His invention, he said, would leave the animal free to lash her tail as much as desired "and all evil effects are overcome." (Figure 9.8)

The face mask for cows patented in 1963 by Henry George O'Hare, Jr. of Amboy, Illinois keeps away annoying insects. Made of hard plastic, it is held in place by a strap under the jaw and has a screen through which the animal can see, undisturbed by flies. In use, according to Mr. O'Hare, it is pleasant for the cow to wear and insures placidity and continued milk production. (Figure 9.9)

Frank M. Kaneski of Duluth, Minnesota found that the tail was in one way a nuisance to the cow herself. He reported in 1952 that "the only useful purpose known for a cow's tail is to keep flies off of her and there are not flies during the winter months." When she lay down, he said, the tail got dirty and when she switched it around she transferred the dirt to her flanks. He immobilized the tail all winter with a ring through the loose skin and a clamp that gripped some of the hair on the cow's tail. In summer, the tail could be released for swinging at flies. (Figure 9.10)

Fig. 9.11:
Back up when you get a shock.
(Patent 1,411,312)

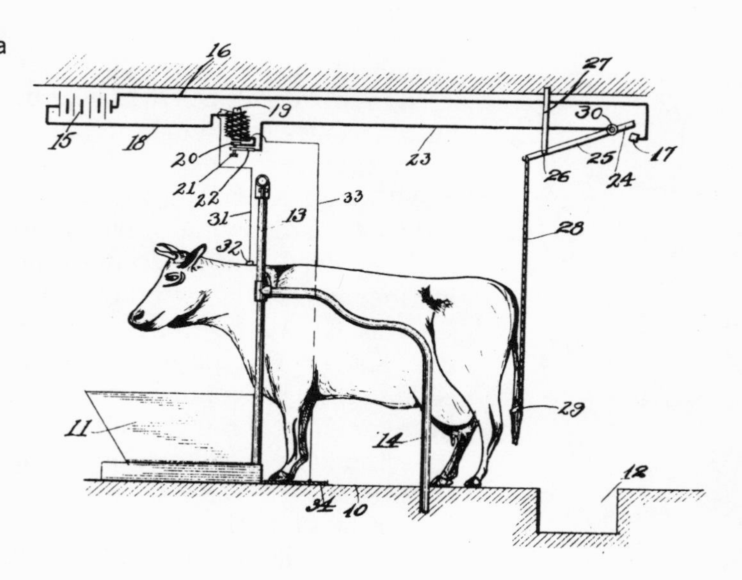

Much earlier, a Wisconsin inventor had wired a cow's tail to force her to move to a position over a gutter when she was about to release droppings. Elmer Swensen of Valders said in his 1922 patent that immediately prior to evacuation the animal would naturally raise its tail. This would close a switch, electrifying a mat upon which the cow's front feet rested and causing her to step backward the proper distance. (Figure 9.11)

Dental care has been available for fortunate bovines. Caps for teeth were patented in 1962 by A. Rood Menter, a rancher at Sedgwick, Colorado and his dentist, Dr. Ward C. Newcomb of Chappell, Nebraska. A bovine dental plate or restoration to cover a cow's lower jaw was contributed two years later by Mr. Menter. (She has no upper front teeth but chews grass against a gummy pad.)

Bulls like to butt, not only each other but such objects as doors, gates and fences. The bull control patented in 1967 by Leonard Pemberton of Harrison, Arkansas discourages the practice by placing on the animal's forehead a pair of wheels provided with spurs. When the bull butts, the spurs dig into him. (Figure 9.12) Another pain-inducer gets into operation when the bull tries to go through or under a wire fence.

Fig. 9.12:
Spurs on forehead stop butting.
(Patent 3,323,493)

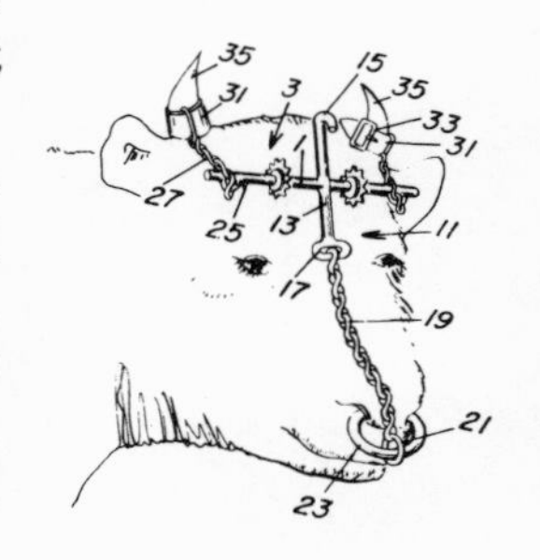

One training aid—not for cattle but for the humans who breed them—is a replica of a cow's hindquarters. The teaching device helps to school technicians and herd managers in the artificial insemination of cattle, an art of great importance in raising both beef and dairy cattle.

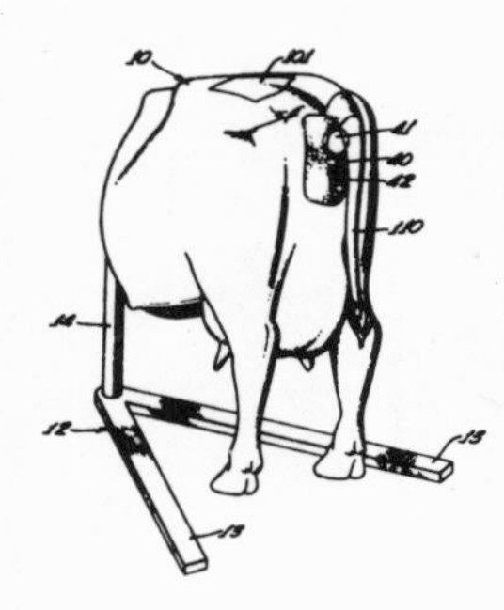

Fig. 9.13:
Artificial rump for student use.
(Patent 3,309,791)

Although a serious tool for the industry, the apparatus is somewhat startling to the lay observer. It is a life-sized reproduction of a cow's rear end, made of fiber glass with windows in one side. Patented in 1967 by Gerald L. Kelley and Roy U. Selover, Jr. for the Kelver Company of Modesto, California, the Kelver Kow drew purchase orders from Purdue and a number of other universities and was approved for distribution abroad under the A.I.D. program. (Figure 9.13)

A related artificial animal, but one that doesn't stand still, is the mechanical calf—used in training ranch horses to herd cattle. The creature, patented in 1967 by Lee R. Harris of Brenham, Texas, looks like a live, 600-pound calf. Under remote radio control, it confronts a horse and rider and maneuvers about. There are three rubber-tired wheels and each of the rear two has its own battery-powered motor. To make the calf turn, the operator reverses one wheel or the other. Calf-trained horses won honors and in a sense the calf won its own, as the deluxe model sold for $2,490. (Figure 9.14)

Fig. 9.14:
Mechanical calf trains horses.
(Patent 3,303,821)

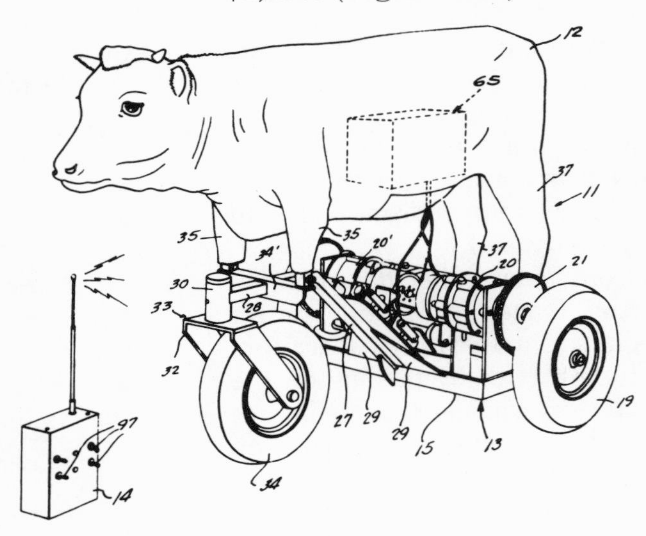

Fig. 9.15:
Cow-shaped refrigerator gives milk.
(Patent 611,653)

For the delivery of milk in city streets, a refrigerator shaped like a cow was patented in 1898. The cow was to be mounted on a truck, with internal milk tanks and storage space for ice. The inventor, David C. Standiford of Baltimore, planned that sweet milk or buttermilk could be drawn when pressure was applied to push-rods in the proper teats. He was sure his portable refrigerator would attract considerable attention by its realistic appearance and the simulated milking. (Figure 9.15)

From Iowa comes an artificial sow, patented in 1964 by Alfred W. and Cora Lee Brown of Ankeny. Their automatic suckling pig feeder is described as the size and shape of a brood sow, having a lighted eye and emitting the recorded sounds of a nursing animal.

The mechanical mother, laid on its side, presents two rows of nipples, usually 14 in number. A clock in the head times a 30-minute operating cycle with the eye periodically turning red, warm milk being pumped and a tape recorder playing through a speaker. The Browns have reported effective field tests and reduction in the mortality rate of baby pigs. The tank holds five gallons. (Figure 9.16)

Fig. 9.16:
Artificial sow feeds hungry piggy.
(Patent 3,122,130)

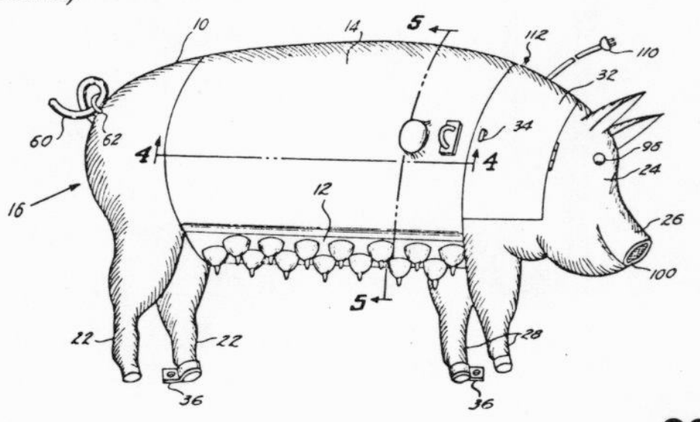

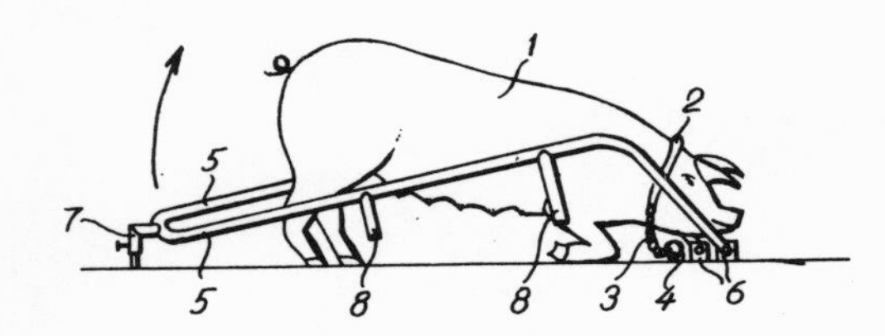

Fig. 9.17:
Now she can't crush her porklings.
(Patent 3,166,049)

Fig. 9.18:
Slapper and shocker moves cattle along.
(Patent 3,227,362)

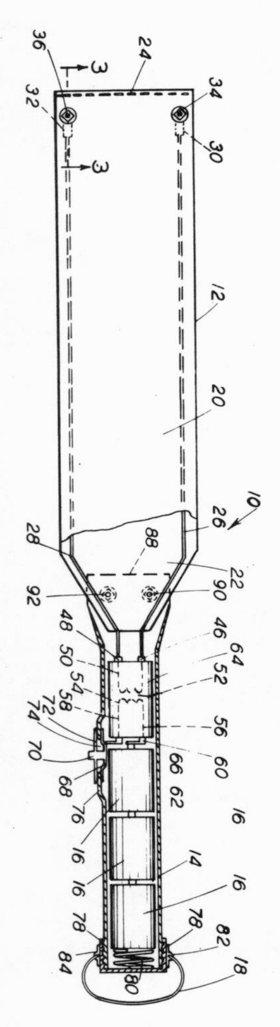

As the Browns noted, baby pigs are sometimes crushed by their natural mothers. A Swedish inventor, Liss Axel Lundin of Hjorted Station, got a 1965 patent entitled "Device for Preventing a Sow from Lying Her Porklings to Death." He explained that a sow normally lays herself down by bending her forelegs and then her hindlegs so that she rests on her belly, after which she rolls over to one side, sometimes catching helpless members of the litter. He offered a framework with a pair of shafts between which the sow could be tethered and which would keep her from rolling over. (Figure 9.17)

Patentees have offered rural residents a variety of intriguing innovations over the years. Among the examples:

- A pump enabling any creature to draw drinking water from a well or cistern by its own weight. When the fowl or animal walks onto a platform, a plunger pumps water into a trough in quantity corresponding to the creature's size. In cold weather a lamp can heat the trough. Sullivan W. Edminston, Hillsboro, Ohio, 1909.
- A waterproof barn roofing composition consisting, in part, of fresh cow manure, chicken manure, mashed half rotten potatoes, mashed half rotten apples and iron water (made by leaving rusty irons in a barrel of rainwater). Frederick Reisig, Sr., Barnesville, Pennsylvania, 1923.
- An electric slapper for driving stubborn hogs, cattle and other livestock. The animals can be struck with the flexible flat blade and simultaneously given electric shocks from contact points protruding from the blade. If the battery runs down, the slapping is still effective. Perry H. Laten, Fremont, Nebraska, 1966. (Figure 9.18)

• Artificial, grasping fingers with metal bones. When poked singly or in groups into the branches of a tree and inflated, the fingers pick fruit. They can also transfer fruit from one container to another. James D. Frost, Porterville, California, 1967 and 1971.

One picture that sticks in the memory after a skim through farm patents is of a man wearing a serious expression, a hat and a beard—holding his arms outstretched in a T. The figure turns out to be a coyote alarm. John S. Barnes of Payette, Idaho explained in 1903 that the scarecrow was to be set up near a sheep pen or corral to keep the coyotes away. About dusk, the rancher would wind the mechanism under its metal breastplate, and it would explode a blank cartridge every 15 minutes all night long. In the morning the rancher—who presumably had been sleeping some distance away—would turn off the detonator and restore quiet. (Figure 9.19)

Fig. 9.19: Scarecrow keeps away the coyotes. (Patent 726,131)

10

Outwitting the Thief
And Shooting around the Corner

Shooting in curves could be both an appealing sport and a logical defense against the enemies of society. In 1870, perhaps on the basis of Civil War experience, James G. Hope of Topeka, Kansas patented his projectile—which he said could be fired in curves with the same accuracy as in straight lines.

Mr. Hope attached to the rear of the bullet a curved tailfin to serve as a rudder. He made optional a flattened point at the projectile's forward end as an additional guide. The inventor cautioned that the gun should be smooth-bore or have straight riflings to keep the bullet from twisting or revolving. When the missile was being loaded, the guides must be placed in exact position for its desired travel, right or left or in any other direction. (Figure 10.1)

Many years later a Philadelphian devised a different way of shooting around the corner. By 1916 trench warfare was a challenge and Jones Wister disclosed a special gun that could be fired over a parapet without endangering the marksman. He attached a curved

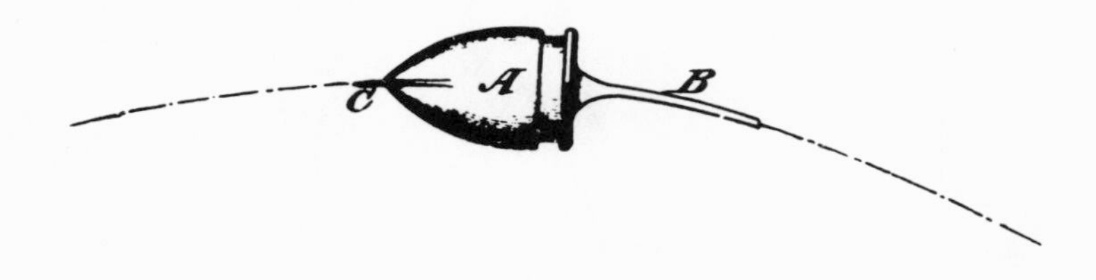

Fig. 10.1:
Bullet that flies in a curve.
(Patent 107,909)

end to the rifle. The infantryman could hold the gun vertical and sight at the enemy by periscope. The bullet went around the turn and took off at the enemy lines.

The invention specifically was a curved extension to be screwed onto the end of the regulation straight barrel. Mr. Wister said the projectile would freely pass through the curved section without binding, although he admitted there would be some frictional resistance. He preferred a round ball or buckshot but thought an elongated bullet could be used. Although he had rifles particularly in mind, he said some machine guns and cannons could be so equipped. (Figure 10.2)

During the Civil War two men in Waterloo, New York invented a combined plow and gun for border localities subject to savage feuds and guerrilla warfare. C. M. French and W. H. Fancher, in their 1862 patent, said their object was to produce a strong light plow "and at the same time to combine in its construction the elements of light ordnance, so that when the occasion occurs it may do valuable service in the capacity of both implements."

The gun barrel, pointed forward, was paralleled by the share buried firmly underground, where it would resist recoil. The inventors thought a projectile of one to three pounds' weight might be accommodated without rendering the invention cumbersome as a plow. "As a means of defense in repelling surprises and skirmishing attacks on those engaged in a peacetime avocation," they said, "it is unrivaled, as it can be immediately brought into action by disengaging the team, and in times of danger may be used in the field, ready charged with its deadly missiles of ball or grape." (Figure 10.3)

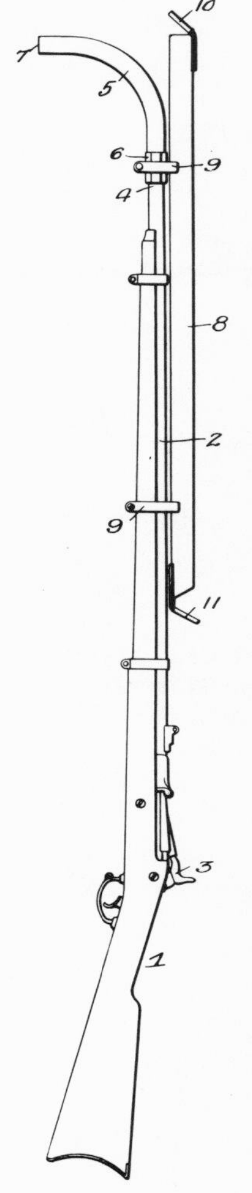

Fig. 10.2:
Gun that shoots around the corner.
(Patent 1,187,218)

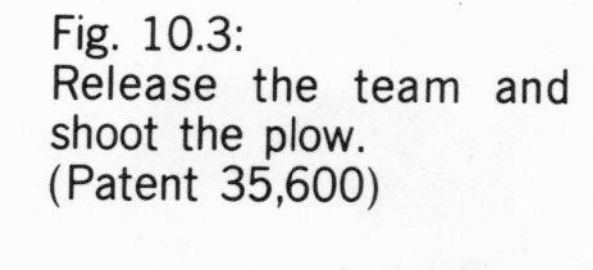
Fig. 10.3:
Release the team and shoot the plow.
(Patent 35,600)

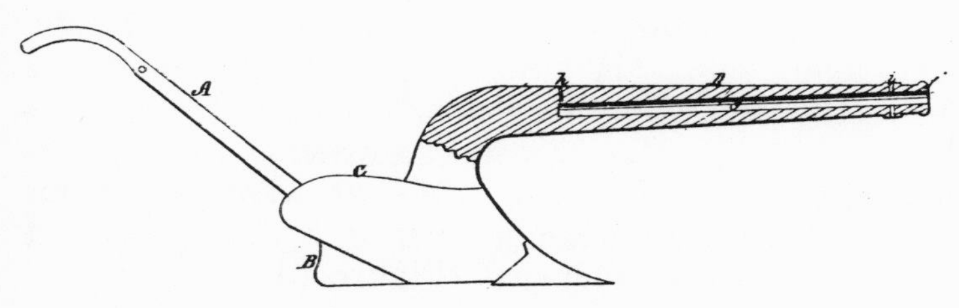

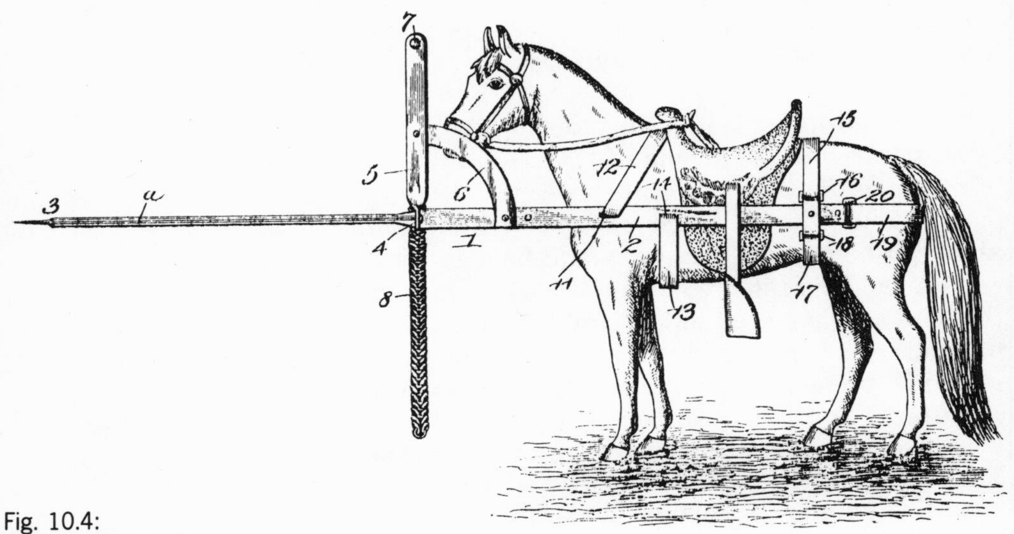

Fig. 10.4:
Cavalry steed aims weapon ahead.
(Patent 636,430)

The invention was disclosed some time before the Civil War ended and before there was talk of beating the nation's swords into plowshares.

Novel cavalry equipment, offering protection for both horse and master, was revealed in 1899 by Franz and Konrad Hieke (described as subjects of the Emperor of Austria-Hungary) who resided in Philadelphia, Pennsylvania. A frame on both sides of the animal—hung from withers and rump—held in front a flat, pointed blade designed to injure an enemy and cause him to move aside during a charge. Chain mail was suspended to shield the horse's forelegs. (Figure 10.4)

Fig. 10.5:
Bicycle thief gets a prick from the seat.
(Patent 650,082)

Another kind of rider was the target of a bicycle attachment patented in 1900—anybody who was likely to steal a two-wheeled vehicle and pedal it away. The miscreant who appropriated a bicycle equipped with the seat devised by Adolph A. Neubauer of Camden, New Jersey was due for a rude and painful shock when he threw himself into the saddle. The seat, Mr. Neubauer said, "is provided with one or more upwardly-projecting needles or pricks." The bicycle's owner could adjust the mechanism so the needles were held securely down while he was aboard but could release them when he parked. A thief would get a deep and sharp surprise. (Figure 10.5)

To trace a thief who moved about on foot, James O'Connell of Bedford, Indiana conceived a footmat that would sprinkle the fellow with fluid and identify him by sight or smell. According to the 1910 patent, the mat was to be placed in a doorway subject to trespass so that when an intruder trod upon it the telltale fluid would be discharged upwardly, tinting the person's garments or imparting a pungent and offensive odor.

A burglar alarm was also designed to spray somebody but for a different purpose. Arnold Zukor of New York disclosed in 1912 equipment to be actuated by the opening of a door or window and to squirt water on the face of a sleeping watchman or home owner. The inventor explained that the usual audible alarm might not be heard by those with defective hearing or too deep in slumber. He added a provision for use in the daytime when the house was vacant: If a window was opened, the outside of the building would be sprayed, indicating to any observer an invasion by unauthorized persons. (Figure 10.6)

Fig. 10.6:
Burglar alarm sprays a sleeper awake.
(Patent 1,046,533)

To thwart the hat thief, Frank P. Snow of Los Angeles placed a sharp-pointed prong in the sweatband. An unauthorized person picking the hat from a rack and putting it on would be painfully jabbed. Mr. Snow explained in 1914 that a man who had merely made a mistake would see he had taken the wrong hat and one who was determined to steal it could not make it wearable without knowing the owner's combination. (Figure 10.7)

A thief snatching a bag filled with valuables might escape in a crowded street but not with the protective means patented in 1925 by August Eimer of New York. Tearing the bag from the carrier's hand would release a belching cloud of white smoke, produced by exposure to the air of such chemicals as titanium tetrachloride. The revealing cloud could be supplemented with tear gas liberated from another container. (Figure 10.8)

Fig. 10.7:
A hat-thief will be jabbed.
(Patent 1,098,691)

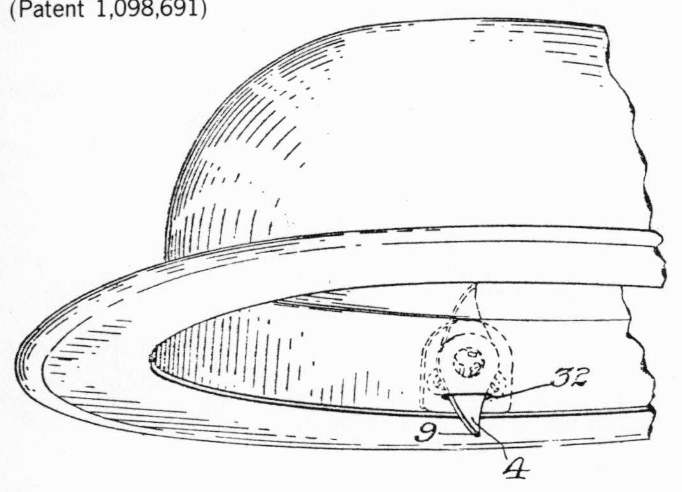

Fig. 10.8:
Snatched valuables release smoke cloud.
(Patent 1,563,176)

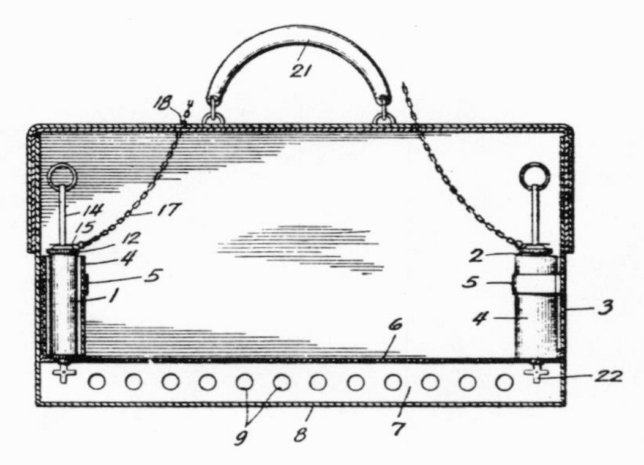

The conventional handbags of lady police officers have telltale bulges that make them undesirable in surveillance work. Velda Alexander of Visalia, California made this point in her 1967 patent for a bag holding a pistol, ammunition, handcuffs, badge and ammunition card and designed not to give itself away. A shoulder strap passes underneath to support the weight. The gun and other articles fit in an insert which can be lifted out as a unit and shifted to another bag.

Hunters and target shooters destroy many traffic and other signs as they drive or walk past. Robert J. Craghead of Brigham City, Utah in 1966 patented a detective camera to make movies of such marksmen for their later identification. A steel housing shaped like a signpost has a sign mounted on its face. When a bullet strikes the sign, the camera takes the shooter's picture, including the license plate number if the offender is in a car. (Figure 10.9)

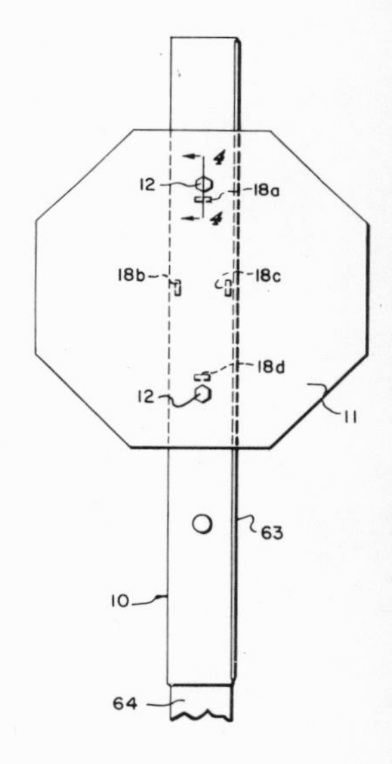

Fig. 10.9:
Shoot this sign and it mugs you.
(Patent 3,244,086)

A deputy county court clerk at Williamson, West Virginia designed a cane to shoot either arrows or pellets. Lee Ellis, Jr. thought the hollow stick would appeal to elderly gentlemen who wanted both a means of defense and something to lean on, and appeal to young sportsmen.

As Mr. Ellis described it in 1962: the device, when not in use, "assumes the outward appearance of an ordinary walking cane but which, upon the removal of a cap from the lower end thereof, may be utilized to project with accuracy either arrows or pellets by means of a spring-actuated plunger." There were a safety trigger and front and rear sights. (Figure 10.10)

Shooting from the hip is fairly common, but not shooting from the belt. What appears to be an ornamental belt buckle opens, according to a 1962 patent, to fire a pistol. The wearer can point the pistol and

Fig. 10.10:
Cane shoots arrows or pellets.
(Patent 3,058,456)

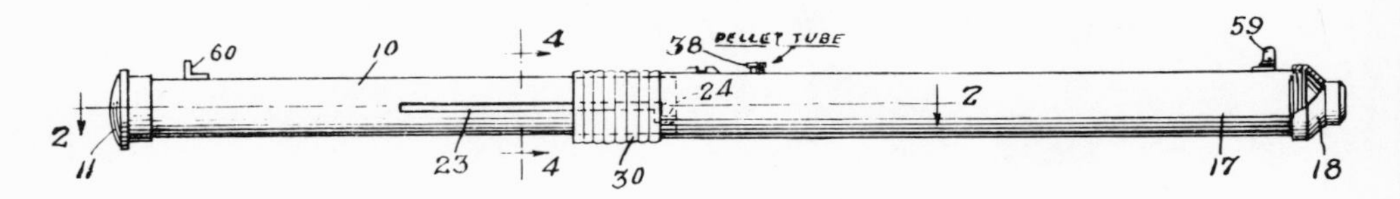

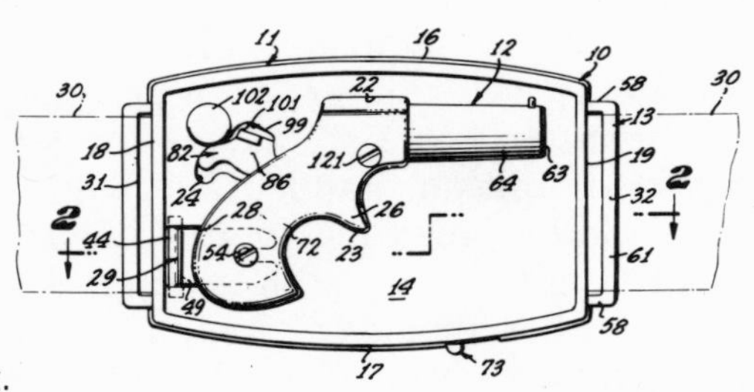

Fig. 10.11:
Belt buckle fires a bullet.
(Patent 3,026,642)

shoot by tensing his muscles and expanding his midriff. John W. Ryan of Bel-Air, California, wanted to provide police with an emergency weapon. If manufactured as a toy, it would enable a boy to feign surrender by raising his arms and amaze his playmates by firing a cap and ejecting a pellet. (Figure 10.11)

Another method of shooting with the hands raised was revealed in 1921 by Harry N. McGrath of San Francisco. In a bank holdup, a cashier could lift his arms and press a small button in his palm to fire an automatic pistol that had been strapped beneath his armpit and hidden by his coat. Mr. McGrath said to make the gun perfectly safe a blank cartridge could be placed in the magazine to be fired first, followed by a ball cartridge. (Figure 10.12)

Cashiers have been offered various ways to foil and trap the bandits that invade banks and stores. As a substitute for firearms, Robert H. Burnell of Washington, D.C. patented in 1953 a burglar trap to enable a cashier—without moving a finger—to knock a bandit off his feet and hold him until the guards or police arrive. With one knee, the cashier releases from under the counter a spring-operated striking bar

Fig. 10.12:
Shooting from the armpit.
(Patent 1,377,015)

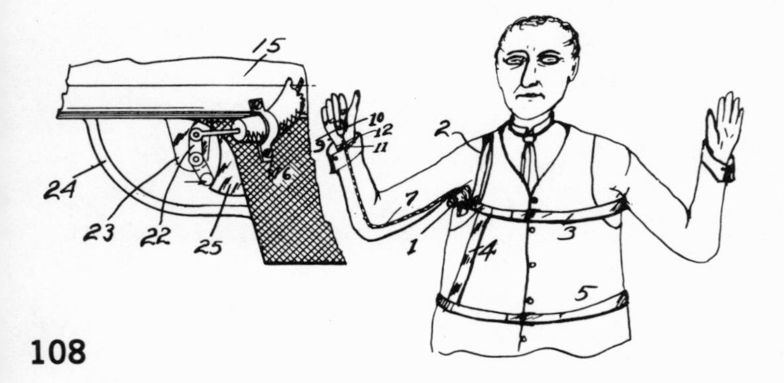

that wallops the robber at about calf level. The bandit is forcibly detained by a series of hooks that clutch his clothing.

A bank bandit trap, as revealed in 1964 by Nadina Billi of Astoria, New York, drops a bullet-proof shield in front of a teller when he presses a button and puts a metal net around the robber. The mechanism sounds a holdup alarm and automatically expels tear gas inside the net. Another trap, as patented in 1967 by Louis A. Turano of Atlantic Highlands, New Jersey is intended to drop a holdup man through the floor and keep him in custody. One button opens the trap doors in front of the teller's cage; another starts a movie camera and tape recorder to keep an account of what goes on. An observation window allows the prisoner to be watched and be sprayed with tear gas.

John Walter Fisher of Woking, England patented in 1964 a bank messenger's bag capable of vigorous self-defense. When seized by a thief, it grabs his hand, blows a police whistle and fires a blank cartridge. The bag also extends three eight-foot telescoping arms to prevent his exit through a door or window or his entrance into a get-away car. To avoid injuring passersby, the arms have rubber buffers on their tips.

Packets of false currency, which can be handed bank robbers, spray them with dye or tear gas. George S. Harner of Garden City, New York patented the dye dispenser in 1962. As the teller passes it over the counter—with bundles of real bills—he presses a button and it explodes during the getaway, also sounding an alarm. Three years later James L. Martin of Atlanta, Georgia patented two forms: one fires a tear gas gun when the bandit picks it up from the counter and the other—thrown into his bag with real money—explodes when he reaches his car. (Figure 10.13)

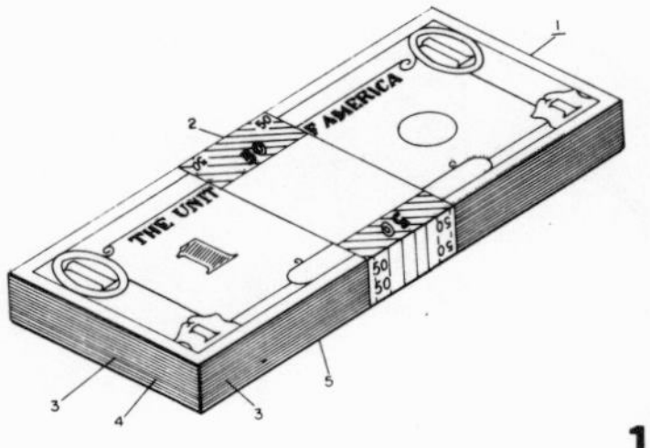

Fig. 10.13:
Fake money fires on a bandit.
(Patent 3,174,245)

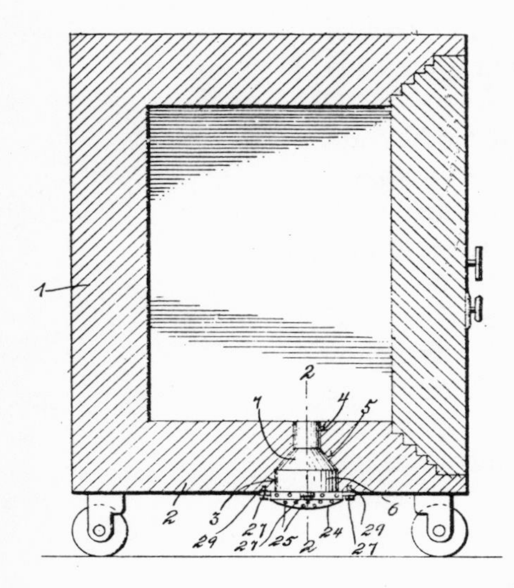

Fig. 10.14:
Safety valve foils safe-cracker.
(Patent 1,129,016)

To frustrate safe-blowers, William H. Ritchie of Patchogue, New York thought up a safety valve. His 1915 patent shows a vent in the bottom of a safe—with a plug that opens if explosives are set off inside—to allow the gases to escape without blowing open the door. Mr. Ritchie's arrangement, he says, protects the safe's contents from fire and prevents tampering through the hole. (Figure 10.14)

A security device that must have strong appeal for bankers is a one-man, indoor, motorized tank to be occupied by the watchman. The vehicle, six or eight feet high and pierced with peepholes and gun openings, has mechanical arms that can grab a thief and hold him till help arrives. The inventor, Stanley Valinski

Fig. 10.15:
Tank grabs and holds a burglar.
(Patent 1,392,095)

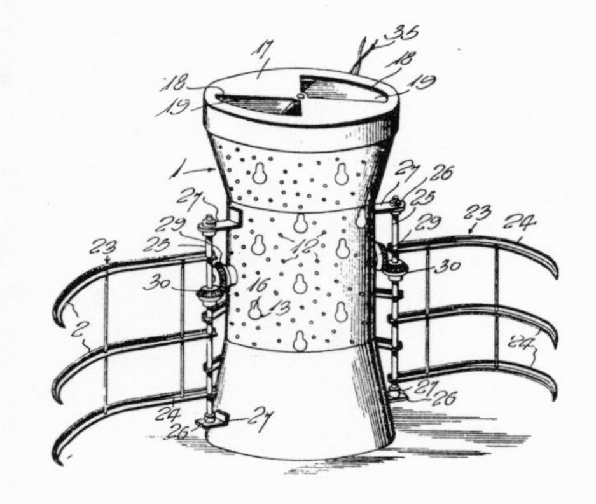

of Homestead, Pennsylvania, provides an electric wall plug through which the watchman can signal the police until he has to pull the plug and move to another part of the bank. "It will be seen," says Mr. Valinski, in his 1921 patent, "that I have devised a device which will enable me to catch, chase and hold burglars or bandits with extreme safety to myself." (Figure 10.15)

Once suspects are in custody, they face other patented equipment. There is, for instance, the apparatus for obtaining and recording criminal confessions. As revealed in 1930 by Helen Adelaide Shelby of Oakland, California the apparatus induces admissions of guilt by use of an apparition—a skeleton with blinking, lighted eyes, a translucent astral outer body and a diaphanous veiling. Amid this supernatural atmosphere, an examiner in an adjoining cubicle asks questions and the suspect's expression and replies are recorded by camera and tape for later use. (Figure 10.16)

In courts of law, officials have been offered such things as a bullet-proof table top and a container for liquor seized in a speakeasy. The armored table top—intended for protection of members of a court assembly against assassination—was revealed in 1912 by Rudolf Lux of Wunschelburg, Germany. The shield could be raised to the vertical by anyone seated

Fig. 10.16: Ghostly skeleton gets confessions. (Patent 1,749,090)

around the table. The prohibition drink collector, patented in 1930 by Robert G. Tetro of Grand Rapids, Michigan consists of a bulb that fits in the agent's pocket and a tube that runs down inside his sleeve. When the bulb is squeezed and released, liquor from a glass on the bar is sucked in without alerting the bartender and can serve as evidence in court. "The construction described," says Mr. Tetro, "is very practical and has been used to a considerable extent, proving its value." (Figure 10.17)

An electric gun patented in 1906 by Samuel T. Foster, Jr. of Victoria, Mexico may have caused some police concern—as it was described as making no noise or smoke. A magnetic projectile was to be drawn through the barrel by a series of coils at increasing speed until it flew out the muzzle. Power was to come from a battery or dynamo.

As World War I threatened to engulf the United States, there was a proposal for magnetic underwater defense. Louis Schramm, Jr. of Armiger, Maryland patented in 1915 an arrangement of powerful electromagnets on the sides of a ship to attract submarines and electrify them, killing their crews or rendering them helpless. When a submarine was captured in this way, lights would flash and bells would sound aboard the ship to announce the triumph. (Figure 10.18)

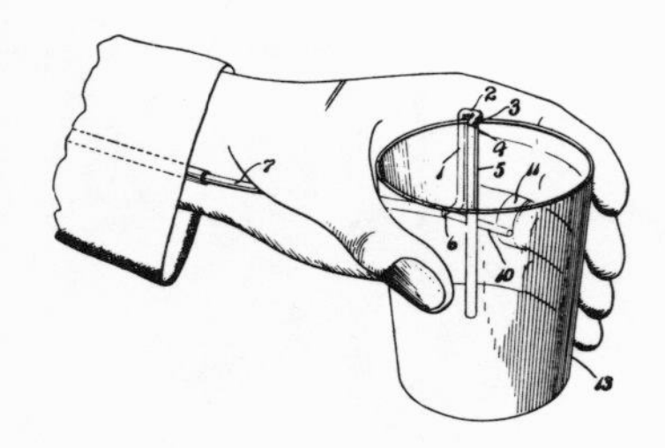

Fig. 10.17:
Suck speakeasy liquor up your sleeve.
(Patent 1,767,820)

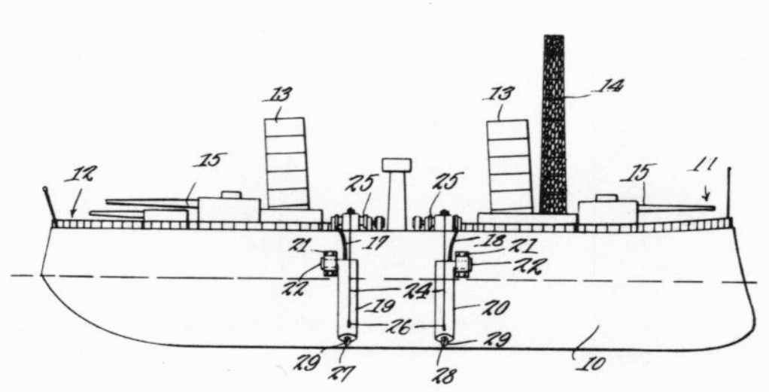

Fig. 10.18:
Magnets capture submarines.
(Patent 1,143,233)

This chapter on domestic and national security will be brought to a close with two minor items relating to the sea. Hannah Rosenblatt, an American who was living in Manila, patented in 1923 a bottle to carry messages across the waves. To attract attention, its cork is topped by a bell hanging from a support shaped like a question mark. "The peculiar shape of the support in combination with the bell," she said, "assures the attraction of attention which is a very important feature of my invention." (Figure 10.19) And Basil Mizerak, while he was at sea in the Navy, invented undershorts with a secret money pocket made of plastic. He said in 1964 that the garment offered maximum security for the funds of sleeping seamen and soldiers.

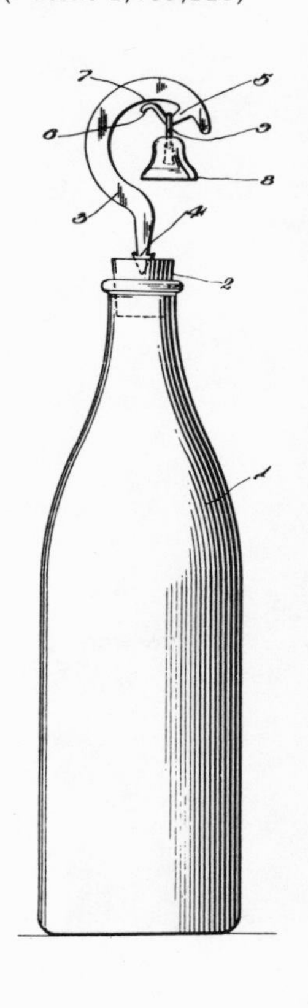

Fig. 10.19:
No question, it's a message bottle.
(Patent 1,469,110)

11

The Healing Arts
And Comfort in the Grave

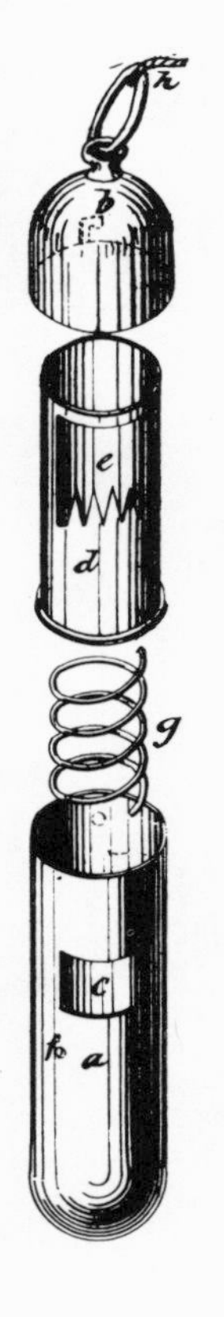

Fig. 11.1:
Swallow a tapeworm trap.
(Patent 11,942)

Medical science has surely progressed since 1854, but at least two inventions recorded in that year are still of interest. One is the trap for removing tapeworms from the stomach and intestines patented by Alpheus Myers, M.D. who practiced in Logansport, Indiana. Dr. Myers described his invention as a trap that is baited, attached to a string and swallowed by the patient after a fast of suitable duration to make the worm hungry.

As the patent explains, the worm seizes the bait and its head is caught in the trap, which is then withdrawn from the patient's stomach by the string which has been left hanging from his mouth, dragging after it the whole length of the worm.

The trap consists of a cylinder of gold, platinum or other metal, about three-quarters of an inch long and a quarter inch in diameter. The bait may be "any nutritious substance." When the worm sticks its head in through a hole, it releases a spring and is caught behind the head. Dr. Myers cautions that the spring must be only strong enough to hold the worm and not strong enough to cut its head off. (Figure 11.1)

A patent granted in the same year to George W. Griswold of Carbondale, Pennsylvania shows how a leg should be sawed off. Mr. (or Dr.) Griswold's surgical instrument, used to assist in amputating limbs, reminds a layman of the clamp that a basement

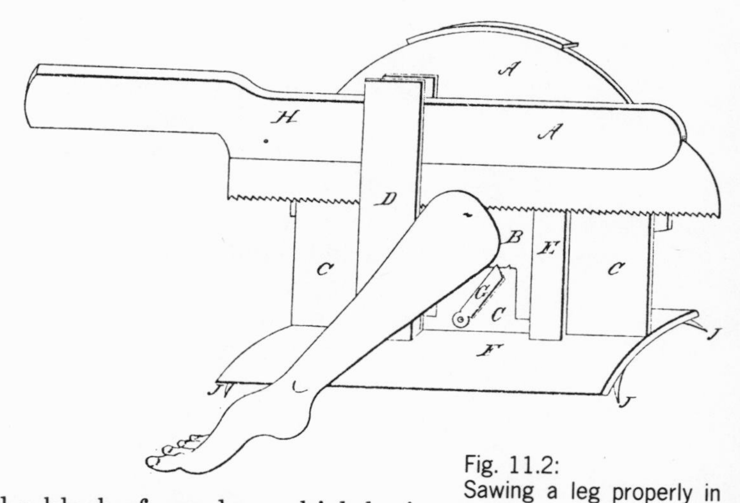

Fig. 11.2:
Sawing a leg properly in two.
(Patent 10,435)

carpenter uses to hold a block of wood on which he is working. It comes into play after the flesh is removed from the limb to be amputated. The surgeon places the bone across the instrument and pushes the apparatus up toward the body to lay bare as much bone as he judges desirable. Then, holding a piece of the equipment down against the bone with one hand, he saws with the other hand. The inventor provides a guide for the saw. In this he says, "the saw moves with certainty and cannot catch, as is the case when the bone is loose, in the ordinary mode of amputation." (Figure 11.2)

Fanny W. Paul of New York offered as a means of inducing sleep a collar that applies pressure to the arteries and veins of the neck, "restoring quiet to the brain in persons suffering from wakefulness or insomnia." According to Ms. Paul's 1885 patent, there is inside the band of the collar a padded spring that the user presses against the flesh below the jaw and about midway between the ear and chin.

Fig. 11.3:
Choke your blood supply and sleep.
(Patent 313,516)

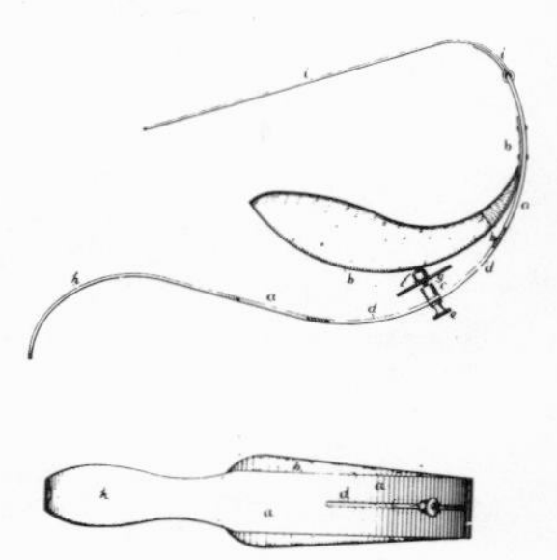

After the undesirable flow of blood is controlled, she says, the nervous system becomes soothed and quieted and sleep follows almost immediately. She reports the result to be attainable in from ten to twenty minutes with no ill effects of any kind. A thumbscrew adapts the device to necks of different sizes. (Figure 11.3)

A medical compound for piles, patented in 1888 by Batist Motea of Burson, California, consists of "common gunpowder, of finest grain, one quart; table salt, one quart; cider vinegar, six quarts." The ingredients are to be thoroughly mixed and the product bottled for sale and use. It is said that if adults take three tablespoons at night, the piles are speedily relieved and prevented. Mr. Motea signed his application with an X and gave the patent rights to William R. Gaylord, one of the witnesses to his signature.

Various kinds of electrical treatment obviously had their appeal in the late nineteenth century. Hercules Sanche of New Orleans used electrical connections to the body, but not to carry current of the usual kind or to transfer heat. According to his 1897 patent, a plate is attached to the person and wires run from it to something that is hot or cold. An attractive drawing shows a lady lying on a bed with a pad on one ankle and a connection running to metal cylinders of coiled wire in a bowl of ice water. Presumably she is being treated for nervousness by a change in her polarity. Besides ice water, the source may be moist earth in which a metal anchor is buried. Says Mr. Sanche:

Fig. 11.4:
Changing a lady's polarity.
(Patent 587,237)

> It is not pretended that there is any flow of current or any of the phenomena of dynamic electricity manifest in this apparatus but only a charging of the body with a certain magnetic polarity, the effects of which upon the system are remarkable in stimulating the system to throw off disease. The cold or earth connection I find almost universally applicable to allay nervousness and stimulate the system by counteracting one polarity of the body.

Mr. Sanche assigned his patent rights to the Animarium Company, of New York. (Figure 11.4)

Ordinary current (strength not indicated) is to be used for the electric extraction of mineral, animal or vegetable poisons from the body by a method patented in 1898. John Bunyan Campbell of Cincinnati placed the patient in a chair with the positive electrode applied preferably to the back of his neck and the negative electrode to his bare feet. The poison is drawn to the negative electrode, but Mr. Campbell says, "For vegetable poisons I employ a vegetable receiver instead of a mineral or copper one and for animal poisons I use an animal receiver, such as raw meat." He adds, "From six to eight treatments of a half an hour each in duration will generally extract all of the poison of whatever kind it may be and the copper plate will show as bright and clear as it was at first." (Figure 11.5)

Static electricity sent through a doctor (or other operator) can be passed to a patient along with

Fig. 11.5:
Poison runs out of his feet.
(Patent 606,887)

vaporized medicine. The principal means is an electrode with a cup on top and a handle at one end, as patented in 1901 by Henry E. Waite of New York.

An influence-machine or static electricity producer (about the size of a large television set) has its negative pole connected to a stand on which the patient sits. The positive pole, grounded on the floor, sends current through the doctor. When he fills the electrode's cup with the desired medicine and starts the machine, static electricity passes through his hand, and the patient gets both current and vapor. "Thus," says Mr. Waite, "the use of my electrode makes it possible to administer mild applications of static electricity to sensitive patients in conjunction with medication." (Figure 11.6)

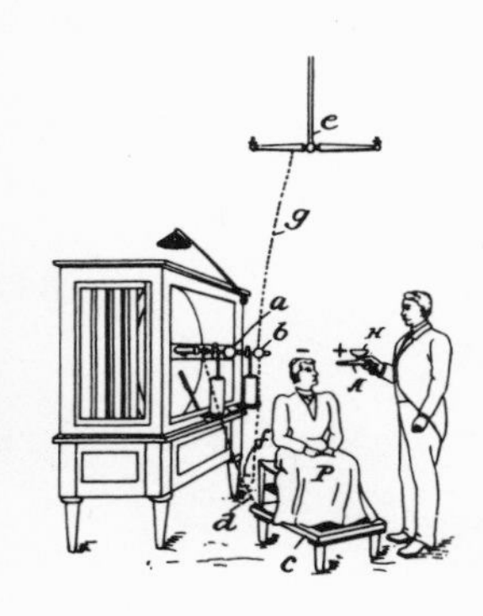

Fig. 11.6:
Patient gets static and medicine.
(Patent 672,047)

By 1915 an Englishman had perfected a method of extracting lead or other metal from the body and collecting it in a vessel on the table in front of the patient. Thomas Maltby Clague of Newcastle-upon-Tyne had as his object the cure and prevention of metallic poisoning. As illustrated, the process is to place the patient's feet in an insulated vessel of water mixed with common salt and his hands in another receptacle holding plain water. Current at about 20 volts is passed through the patient from a positive electrode in the footbath and the lead, or other metal, is deposited in the upper vessel where there is a negative electrode. Besides lead—the inventor says—antimony, arsenic, copper, mercury and silver are easily removed in this way. (Figure 11.7)

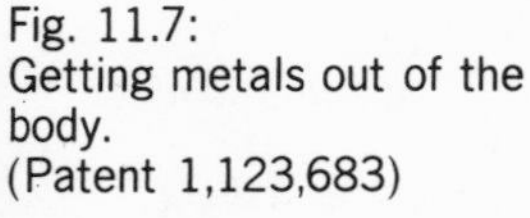

Fig. 11.7:
Getting metals out of the body.
(Patent 1,123,683)

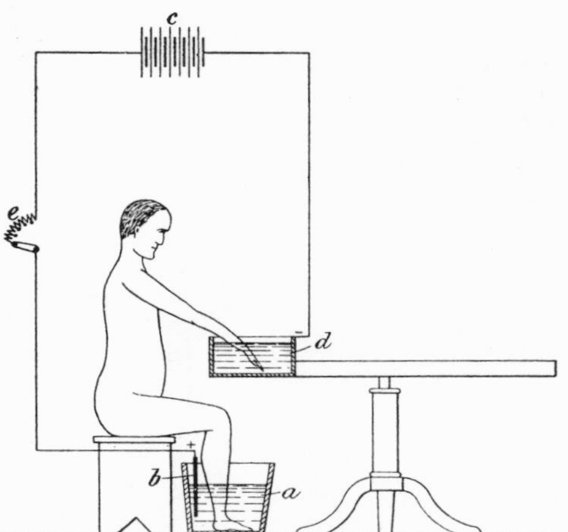

Surprisingly enough, the sexual armor invented by Ellen E. Perkins of Beaver Bay, Minnesota is not for defense against assault but to keep the wearer from playing with himself. In her 1908 patent she calls the practice that the garment is designed to prevent one of the most common causes of insanity, imbecility and feeblemindedness—especially in youth—and equally true of both sexes. Her profession, by implication nursing, had made her very familiar with the subject.

The armor proper is attached to a cloth garment like an armless, short-legged bathing suit. The crotch portion is a rigid metal arch plate with a lock that, when necessary, an attendant can open with a key. Liquid may be passed without unlocking the armor.

"In actual practice," says the inventor, "I have found that an armor or device of the character. . . described, when properly made and fitted to a patient, may be worn with very little, if any, discomfort and that when properly covered by over garments the fact of its application will not be noticeable." (Figure 11.8)

Fig. 11.8:
They call it sexual armor.
(Patent 875,845)

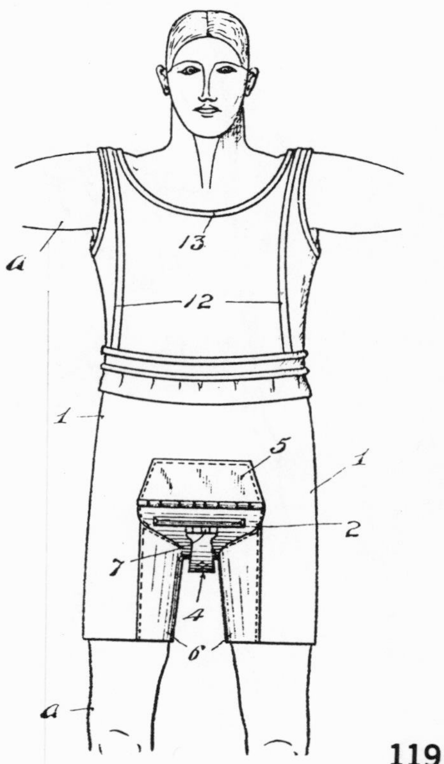

Hats were heavier and more popular 60 years ago than they are today. Bernard H. Nichols of Ravenna, Ohio patented one in 1913 that he designed for the prevention and cure of premature baldness. His hat has depressions in the inner band placed so as not to interfere with the free circulation of blood to the scalp. At the same time, he says, it can be "worn without discomfort and without causing a temporary unseemly marking on the forehead or scalp of the wearer. . . when the hat is removed."

Mr. (or Dr.) Nichols evidently studied many heads and decided that in the majority the four principal branches of the temporal artery were 50, 70 and 100 degrees of arc from the front of the forehead. He arranged his depressions in both sides of the hatband at those points. (Figure 11.9)

Fig. 11.9:
Hat prevents premature baldness.
(Patent 1,062,025)

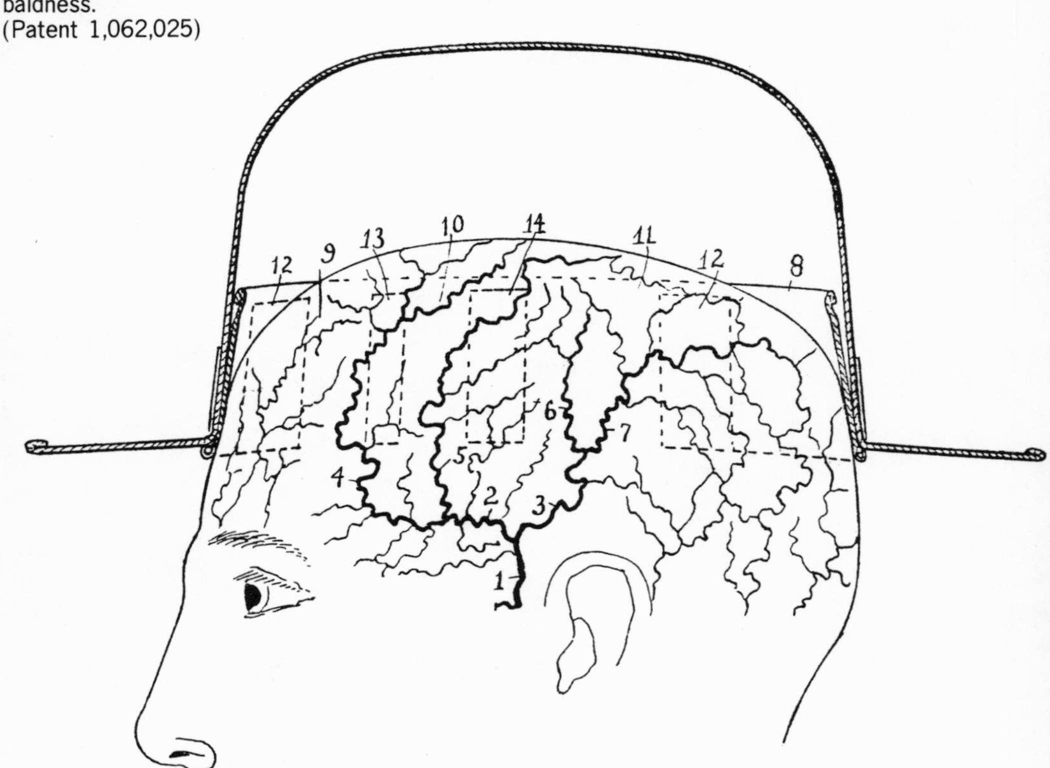

Gases originating in the intestinal canal are offered free escape to the atmosphere through a rectal ventilating deflator recorded in 1918 by Ernest D. Porter of Los Angeles. It is a hollow tube about three inches long with caps and apertures at both ends. The device, Mr. Porter explains, is preferably to be used in the rectum during the night but can be retained during the day, constantly venting the gases and relieving flatulency and constipation. "By making the device of metal," he said, "electric treatment may be combined with the normal function of the device by connecting the same to a source of electrical energy." (Figure 11.10)

A New York mining engineer and his wife patented in 1965 an apparatus for facilitating the birth of a child by centrifugal force. The prospective mother is to be placed on a revolving platform and whirled around under the direction of a gynecologist.

George B. and Charlotte E. Blonsky call their machine the Spinnet. The mother is to lie on a stretcher with her feet toward the turntable rim. Smooth rotation is assured by the use of water ballast and the baby is received in a padded net. Its arrival switches off the motor and rings a bell announcing the event. (Figure 11.11)

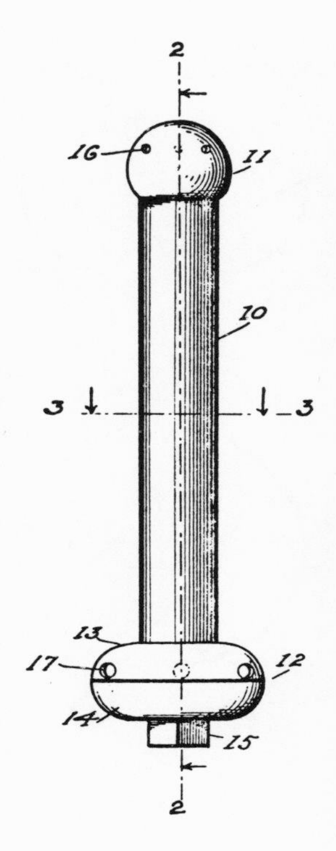

Fig. 11.10:
Gas piping for personal use.
(Patent 1,273,665)

Fig. 11.11:
Centrifugal force brings the baby.
(Patent 3,216,423)

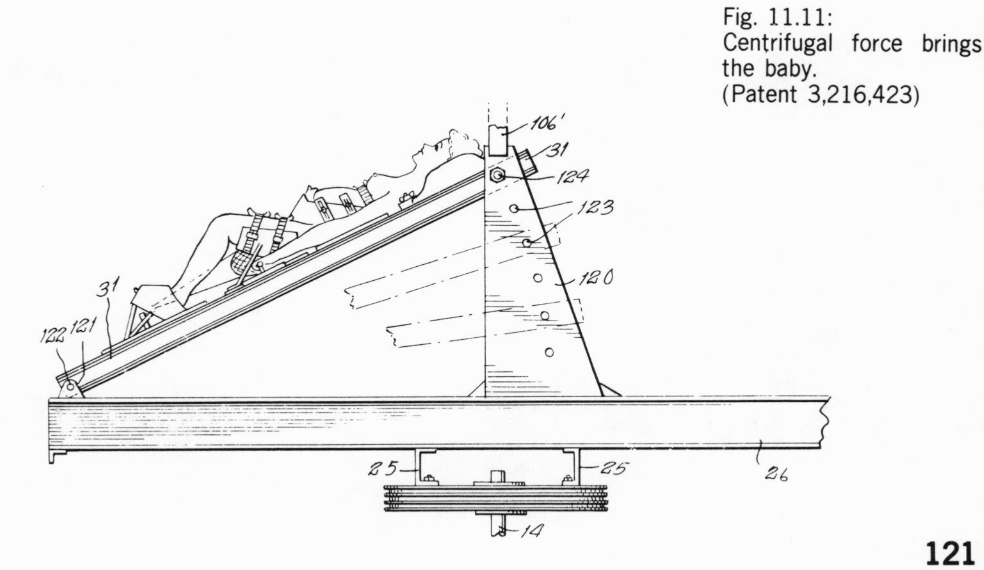

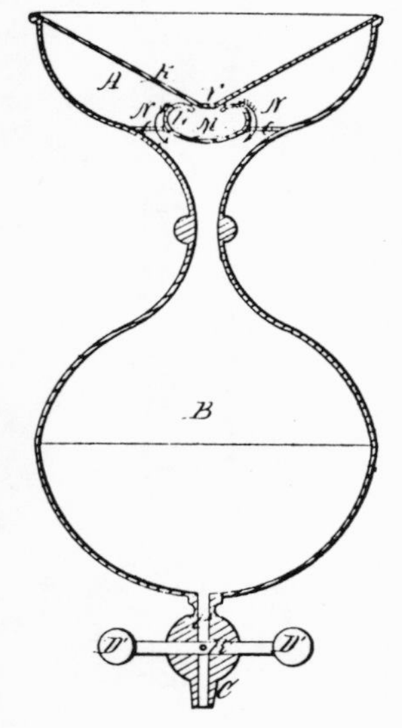

Fig. 11.12:
Spittoon catches gold and silver.
(Patent 51,552)

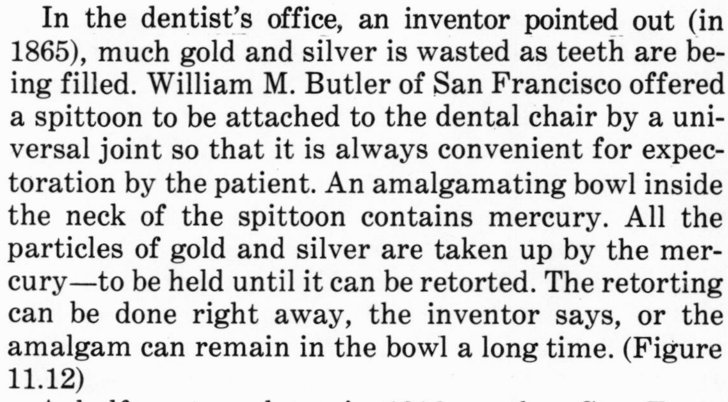

In the dentist's office, an inventor pointed out (in 1865), much gold and silver is wasted as teeth are being filled. William M. Butler of San Francisco offered a spittoon to be attached to the dental chair by a universal joint so that it is always convenient for expectoration by the patient. An amalgamating bowl inside the neck of the spittoon contains mercury. All the particles of gold and silver are taken up by the mercury—to be held until it can be retorted. The retorting can be done right away, the inventor says, or the amalgam can remain in the bowl a long time. (Figure 11.12)

A half century later in 1916 another San Franciscan, Bernard J. Post, commented that permanent proximity of a cuspidor to the arm of a dental chair would subject the more delicate patient to nausea. His invention is a spittoon that can be elevated almost instantly so that the patient can expectorate without having it in full view at all times. The cuspidor rests on a base near the floor until it is needed, when a switch starts the elevating motor and turns on a light. The base may be provided with wheels so that the facility may be moved about for hotel or hospital spitters who might otherwise contaminate the floor. (Figure 11.13)

A much earlier appliance for dental chairs, patented in 1886 by Levi L. Deckard of Middletown, Pennsylvania, administers electricity as an anesthetic. A magneto-electric machine, placed behind the chair, is operated by the dentist with a

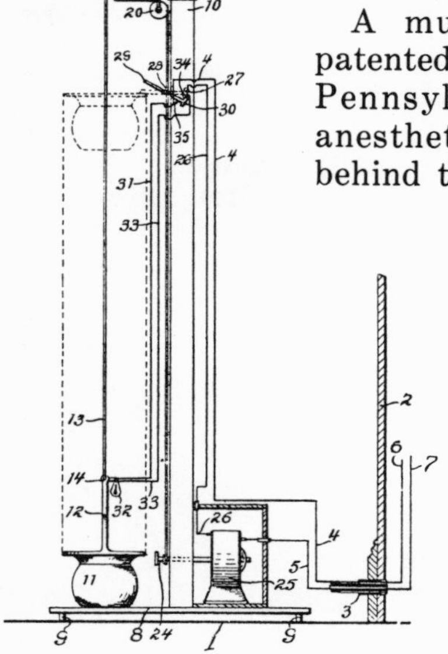

Fig. 11.13:
Lift the spittoon within range.
(Patent 1,193,607)

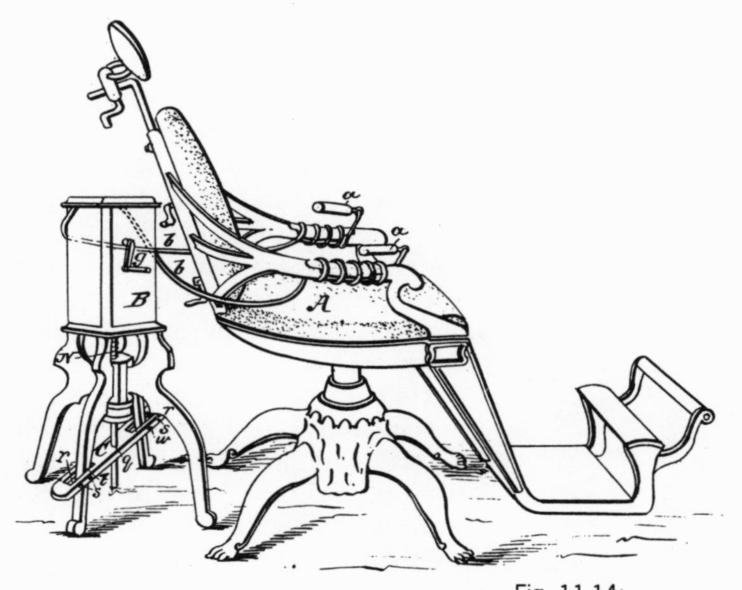

Fig. 11.14:
Electrical dental anesthetic.
(Patent 353,403)

pedal sending current to an electrode on each arm of the chair. Mr. Deckard explains that the current passes through the patient's body when he grasps the arms and the desired shock is produced. (Figure 11.14)

One way a dental patient can signal "ouch" even though his jaws are propped open and his mouth is filled with cotton and machinery comes from Colombia. Max F. Grögl and Arnold Roe of Barranquilla disclosed in 1953 a switch that the occupant of a dental chair can press to signal the practitioner and at the same time disconnect the grinding tool.

What do dental students practice on? An artificial head was offered in 1891 by Howard C. Magnusson of Chicago. As he describes it, the head is made of papier-maché or wood with a metal jaw that can be held closed or open at any angle. The head can be screwed to the back of a dentist's chair in a natural pose. Old human teeth can be fastened in either jaw and removed. "It is obvious," says Mr. Magnusson, "that this invention is an invaluable aid to both professors and students in dental colleges, since in all institutions of this kind there is always a scarcity of patients with their teeth in all the conditions of disease found in a large practice." (Figure 11.15)

Fig. 11.15:
Artificial head for dentists.
(Patent 451,061)

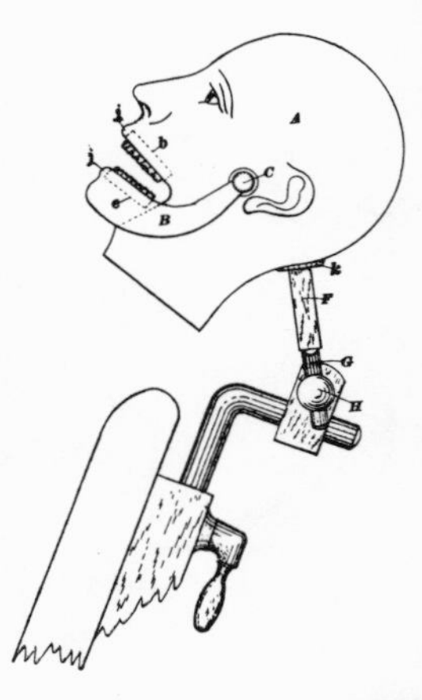

In recent years manikins for dental school instruction in taking X-rays have been made with human skulls imported from Asia. In 1970 Dr. C. Larry Crabtree and two other Public Health Service dentists patented "Dexter" (dental X-ray teaching and training replica). As the use of American skulls has been prohibited by law, those fitted into the manikins are from countries where the vegetarian diet is said to preserve the teeth. They are filled with plastic having the radiation qualities of human tissue and rubber and plastic foam outside simulate face and head. (Figure 11.16)

Considerable ingenuity has been devoted to means of escape or rescue for persons buried while still alive. An 1868 example is the burial-case devised by Franz Vester of Newark, New Jersey. A vertical tube containing a ladder and a cord runs up from a point on the coffin lid directly over the face of the body inside. The lower end of the cord is placed in the person's hand and the upper end is connected to a bell at the top of the tube. "Should a person be interred ere life is extinct," says Mr. Vester, "he can, on recovery of con-

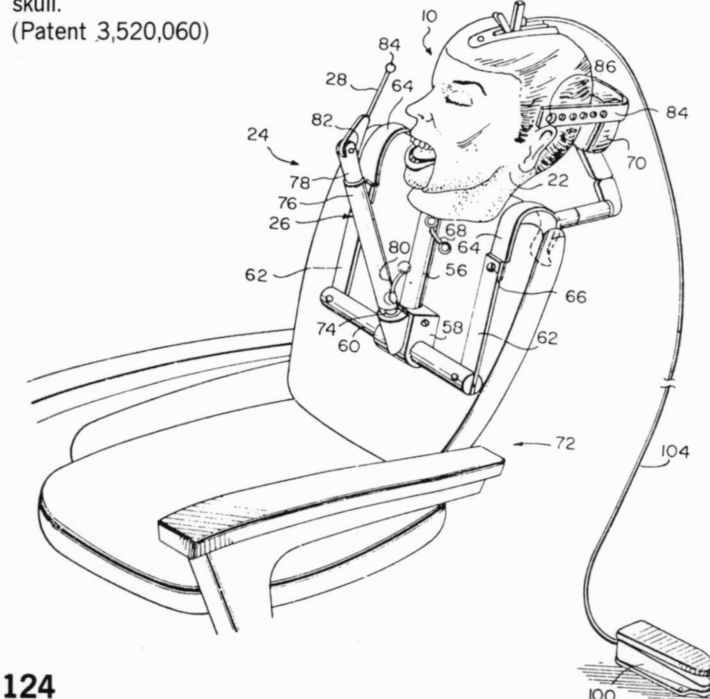

Fig. 11.16:
Manikin with Asian human skull.
(Patent 3,520,060)

sciousness, ascend from the grave and the coffin by the ladder; or, if not able to ascend by said ladder, ring the bell, thereby giving an alarm and thus save himself from premature burial and death."

Nine years later, in 1887, Carl Redl of Vienna, Austria-Hungary added an electric alarm. The coffin may be under earth or in a vault or sarcophagus. By pulling a cord, the buried person can open plates to admit fresh air and connect the battery current to the bell or buzzer. He has to wait for a rescue party, however, as the tube above him is too small for climbing.

An American inventor, William H. White of Topeka, Kansas, in his 1891 version attaches a wire to a ring on the buried subject's finger. A pull starts an annunciator and admits air through a flue. If no signal is made by the annunciator "within a proper time" the upper section of the flue may be removed without letting any dirt fall in or letting any generated gases escape from the coffin. (Figure 11.17)

Fig. 11.17:
Buried person rings for help.
(Patent 455,446)

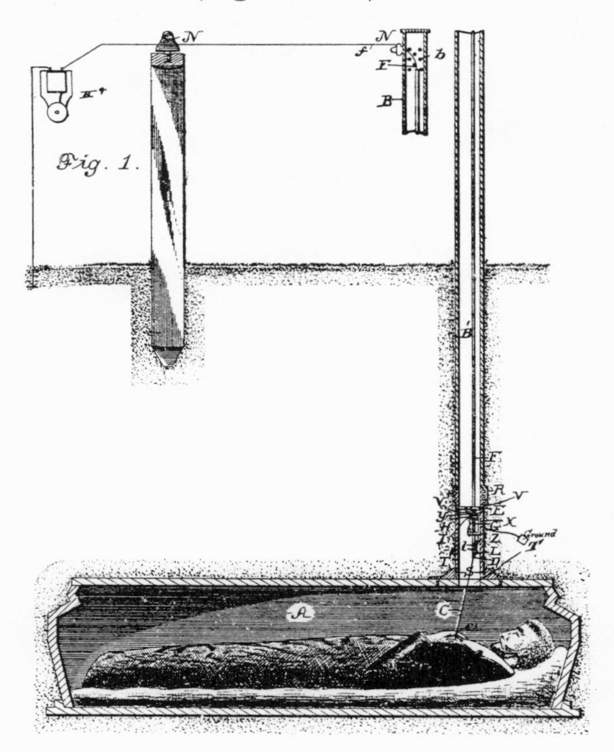

In an 1893 patent a Prussian, Adalbert Kwiatkowski, added equipment intended to give an alarm at the slightest movement of the interred person. A bridle made of soft material rests on the forehead, loose cords pass under the hands and a girdle passes under the body. A pull on any of them displays at the top of the overhead tube a tuft of thread, feathers or hair, "thus warning the outer world that the buried is alive."

Other inventions are intended to enable relatives and friends to look at a body in the grave. Near the foot end of a buried casket George H. Willems of Roanoke, Illinois places a vertical tube, with mirrors at top and bottom. The 1908 patent says that an observer can look into the upper mirror and by turning a rod adjust the lower one and switch on an electric light. If the grave occupant is still alive, fresh air can be admitted. (Figure 11.18)

Jacob Fishman of St. Louis revealed in 1922 what he regarded as an improvement on the above invention. His viewing tube is placed directly above the subject's face and is adjustable in height so that it can be used by either an adult or a child and "eliminates any unsightly aspect in a cemetery." The tube lid can be locked to keep out rain and particles of earth and to insure that unauthorized persons may not indulge their curiosity.

Fig. 11.18:
The interred is viewed by mirror.
(Patent 901,407)

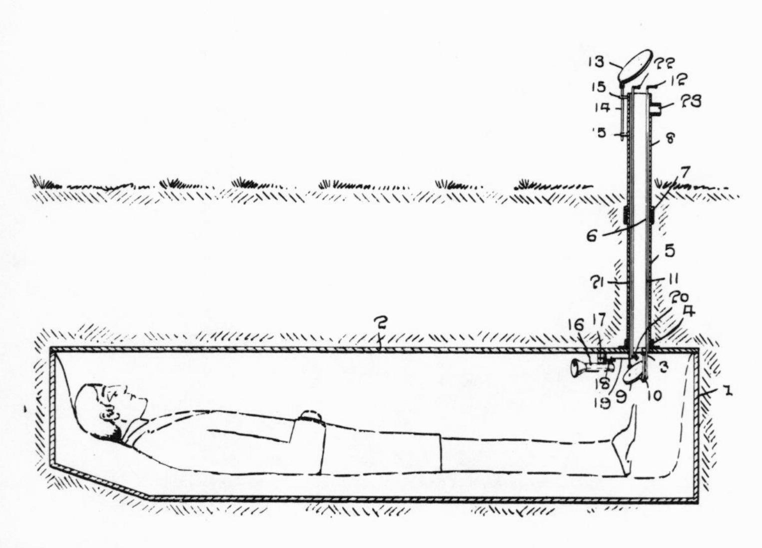

Corpse coolers in various forms were offered in the nineteenth century. As Frederick N. Troll of Baltimore expressed it in 1873, their purpose was "preserving the bodies of deceased persons until it is convenient for their friends to bury them." His improved ice-casket has a rubber lining permitting the air to be exhausted. "The quantum of oxygen necessary to putrefaction being thus removed," the patent says, "and a resupply being prevented by the rubber lining, decomposition is effectually arrested." The ice-chamber is directly over the breast and stomach of the body, "the parts in which decomposition begins."

Highly decorative fences for cemetery lots are described in the 1881 patent granted Jesse Kinney of Detroit. The posts and horizontal rods are of hollow gas or water pipe, with draped chains. The decorations include pots of growing or metallic flowers, clasped hands, crosses, angelic figures, insignia of the Odd Fellows and other lodges and tablets. Tablet inscriptions suggested are "Dick Dead Eye" and "Buttercup." (Figure 11.19)

Fig. 11.19: Ornamental cemetery fence. (Patent 237,024)

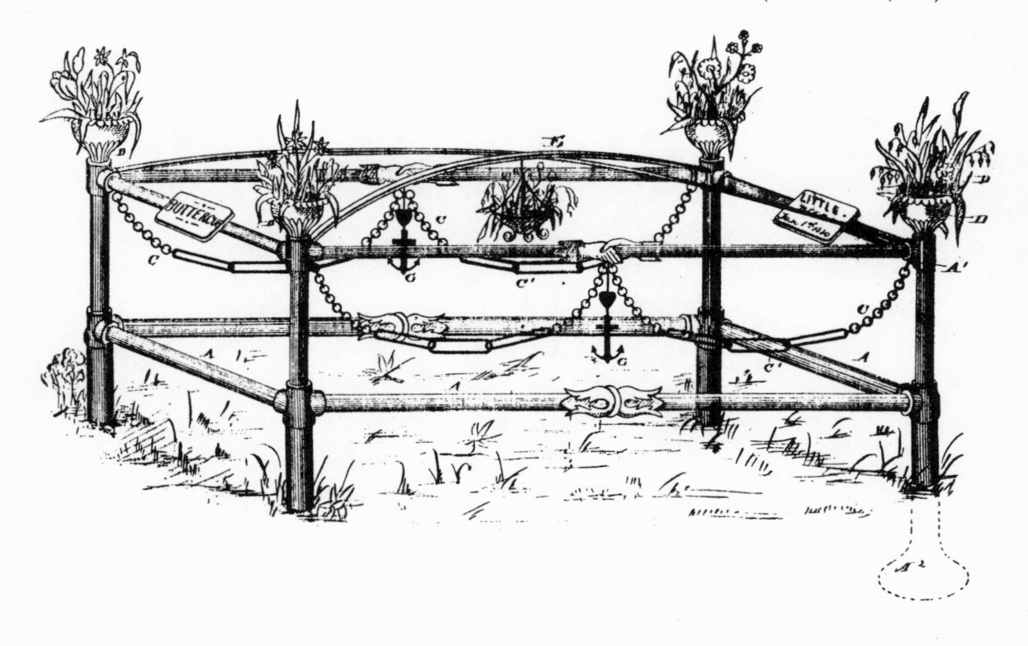

The dead can be preserved in many ways, one of which is being hermetically sealed in a block of transparent glass. Joseph Karwowski, listed in his 1903 patent as a subject of the Czar of Russia and a resident of Herkimer, New York, says the corpse may thus be maintained for an indefinite period in a perfect and life-like condition. The body is first surrounded with a thick layer of sodium silicate or water-glass from which the water is evaporated by dry heat. Then molten glass is applied to the desired thickness. The block may be rectangular or cylindrical. And, Mr. Karwowski adds, a human head alone may be preserved in this manner. (Figure 11.20)

A glass container, but not a solid block, is proposed in the 1910 patent awarded Angelo R. Lerro, an Italian subject residing in Philadelphia. "Many persons," Mr. Lerro says, "object to the ordinary burial casket from reasons of sentiment, because of the unpleasant associations attending the thought of decay resulting from such a disposal of the dead, while others object, for sanitary reasons, to the pollution of the soil and watercourses by the products of putrefaction or decay of a human body interred in a perishable casket." His casket is bell-shaped and encloses the body seated on a stool, which is provided with head

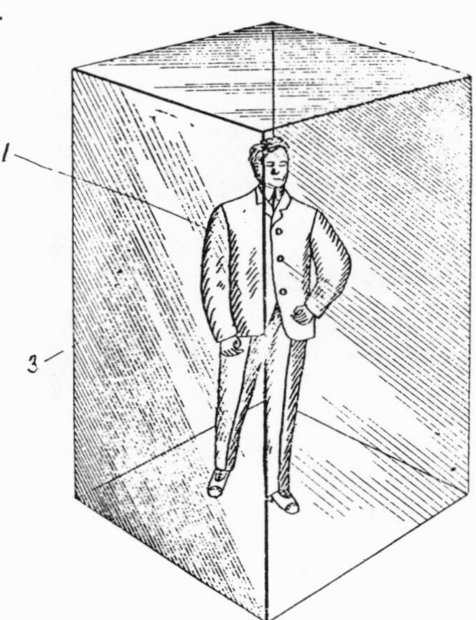

Fig. 11.20:
Buried in a block of glass.
(Patent 748,284)

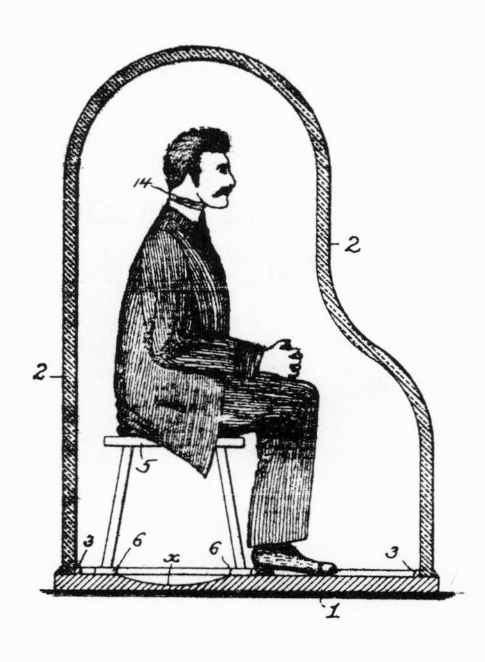

Fig. 11.21:
Seated and awaiting burial.
(Patent 964,439)

and arm-rests. After proper embalming, introduction of a preservative atmosphere and the sealing of the casket, interment may be long delayed. (Figure 11.21)

To save burial space in cities where the expanding population makes it scarce, Ernest E. Bauermeister of Fort Wayne, Indiana urges a vertical casket. As described in 1965, it is an aluminum cylinder with a tapering top that ends in a ring by which it may be lowered. About five vertical caskets will fit in the area required for one of the conventional kind. (Figure 11.22)

Cryoembalming is a possibility. Bodies can be stored indefinitely at extremely low temperature by the method that Richard Pauliukonis of Cleveland protected in 1967. Caskets may be wheeled into hollow storage passages that are surrounded by liquid nitrogen. The bodies, it is said, can be held in a frozen-state condition that, biologically, will remain almost absolutely stable for an indefinite interval of months, years or millenniums.

Much attention has been given to ways and means of preserving corpses. But, on the whole, America seems to be more interested in preserving live and healthy bodies and keeping them productively active in their normal environment.

Fig. 11.22:
Vertical casket saves space.
(Patent 3,188,712)

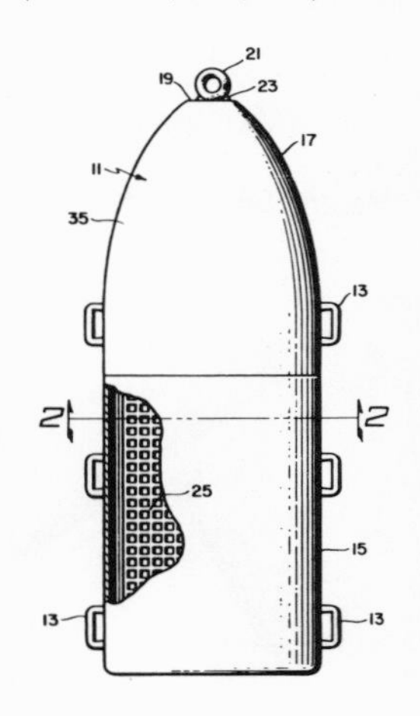

12

Care of the Person

When Your Best Friends Won't Tell

To many men, the morning ceremony they perform before the mirror with razor and shave cream is a daily grind. And a grind it would literally be if they could find in the corner drugstore the device for shaving by abrasion, which was invented by Samuel L. Bligh of Custard's, Pennsylvania. Indeed, any manufacturer would be free to put it on the market today, as Mr. Bligh's 1900 patent expired long ago.

The Bligh beard grinder turns out to be a wooden roller coated with emery which the user holds against his face and rotates at high speed. A belt from the driving wheel of a sewing machine passes over the roller and the shaver provides the power by pedaling. He moves the roller about over his face as the beard is worn away. Mr. Bligh emphasizes that no soap or water is used and the face is perfectly dry. (Figure 12.1)

Fig. 12.1:
Just grind that beard away.
(Patent 646,065)

Two relatively recent inventions pertaining to facial hair are sideburn gauges. Two Connecticut men, John M. Daniel of Bridgeport and Eric A. Hultgren of Fairfield, in 1956 patented a guide to be hung first on one ear and then on the other, while the shaver cuts the sideburns in correct alignment. Another instrument, disclosed in 1957, disregards the ears as points to measure from and employs a spirit level. The gauge devised by Jacob Cohen, a Chicago laundry owner, fits over the top of the head like a telephone headset and has a graduated metal tab pointed down each side. The tabs are set at the desired length and when the barber or individual owner sees the bubble in the middle of the spirit level he knows the gauge is properly centered and starts trimming. (Figure 12.2)

A razor with a radio in the handle informs the user whether he is getting a smooth shave. John H. Worthington of West Lafayette, Indiana explained in 1966 that his device makes it unnecessary for the shaver to check with his fingers for whiskers that have escaped the blade. If any beard remains, there is a noise from a loudspeaker like static, but if the check is smooth the sound is only a soft whisper.

An appliance to be worn in the mouth is called a weight-repulsing and cigarette-reneging appliance. James M. Stubbs of Rockingham, North Carolina, intended the device to discourage smoking and eating by stimulating the flow of saliva and satisfying the craving for food and tobacco. As pictured in his 1965 patent, it fits around the teeth and in the cheeks and is equipped with a tab or button, "for frivolous movement by the tongue." (Figure 12.3)

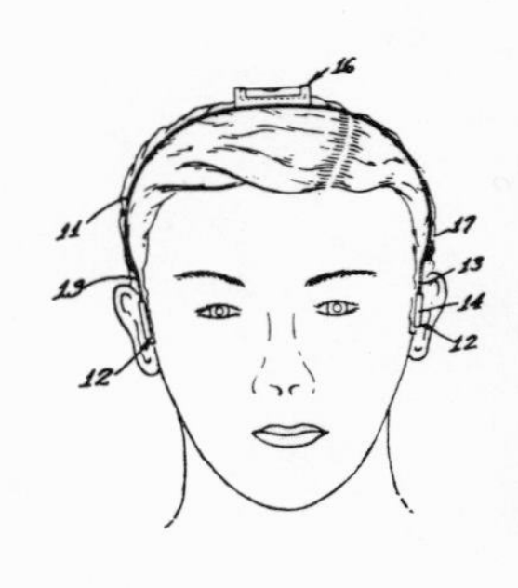

Fig. 12.2:
Trim sideburns with spirit level.
(Patent 2,786,477)

Fig. 12.3:
Eat not, smoke not.
(Patent 3,224,442)

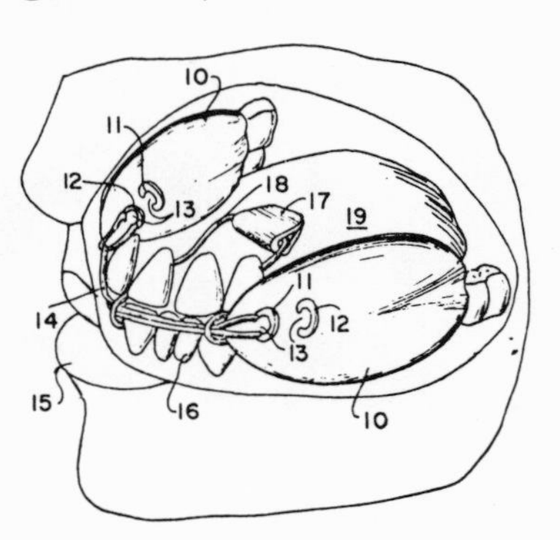

Mustache styles change, as do the means of dealing with them. An Oakland, California woman, May Evans Harrington, described in 1889 her "mustache-guard for attachment to spoons or cups when used in the act of eating soup and other liquid food, or drinking coffee." The spoon or cup is raised to the mouth "and the guard will first pass under the mustache and raise it so that it will not touch the contents of the cup or spoon." (Figure 12.4)

In 1901 Thomas Ferry of Wilmington, Delaware offered a guard for attachment to the mustache itself, rather than to an eating-utensil. "The invention," he said, "relates to mustache-guards designed to hold the mustache away from the lips and to prevent the lodgment of food thereon while eating." It may be conveniently carried in the vest-pocket, he added. The user inserts through his mustache, from below, a series of upwardly-inclined teeth and then straps an elastic tape around the mustache. Prongs on the guard "serve to support the long flowing ends of the mustache, which otherwise might droop down in the way." (Figure 12.5)

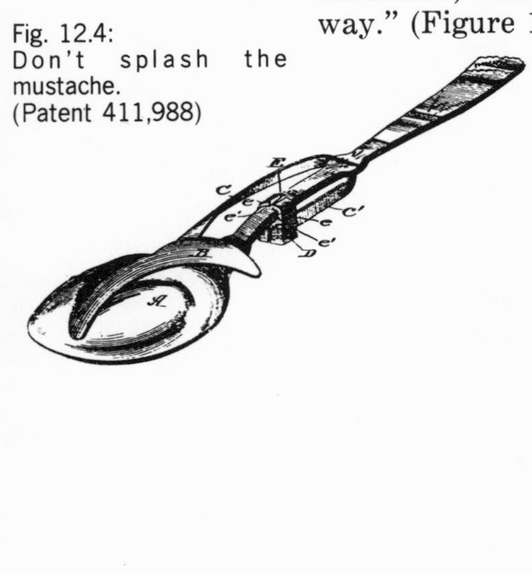

Fig. 12.4:
Don't splash the mustache.
(Patent 411,988)

Fig. 12.5:
Keep it up out of the soup.
(Patent 680,578)

Hair tonics patented over the years call for familiar but sometimes surprising ingredients. Beverley Harris of New Orleans said in 1859 that the bitter apple and gunpowder included in his formula open the pores of the skin and invigorate the roots of the hair. He included also castor oil, bay rum, alcohol and quinine. Just a half century later, Thomas H. Bartlett of Los Angeles said he baked pork and beans for twelve hours to obtain grease, to which he added mutton tallow, olive oil and camphor, scenting the tonic with lavender or rose oil. In the same year Friedrich W. E. Müller of Chicago recommended—especially for bald spots—a composition including ripe black currants, sugar, "best" corn whiskey and port wine.

As a substitute for and improvement upon the conventional haircut, John J. Boax of McKeesport, Pennsylvania proposed in 1951 an apparatus to cause the hair to stand on end and to singe off the tips. He describes a mask that fits over the top of a man's head, not over the face. Exhaust equipment pulls up the hair and electric resistance elements—instead of the conventional scissors—do the trimming. Mr. Boax gives this advice to barbers: "Since the size and shape of the head varies and since different men have different ideas regarding the length of their hair, it is preferred that a set of masks of various sizes and shapes be provided to suit different sizes and shapes of heads as well as to provide for different lengths of hair." (Figure 12.6)

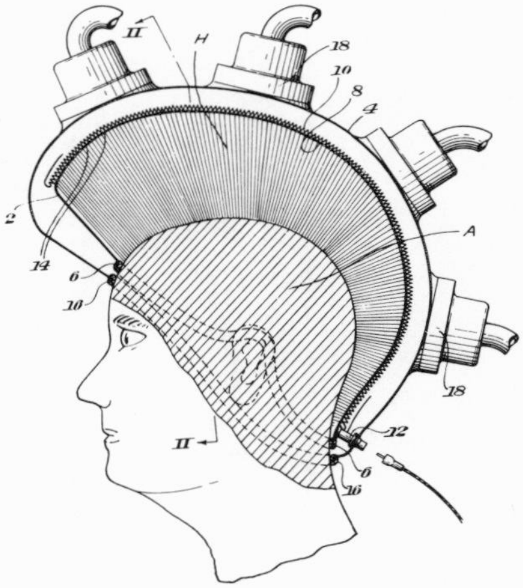

Fig. 12.6:
Hair, stand up and be trimmed.
(Patent 2,577,839)

Another suction device is a dandruff-removing comb that can be attached to a vacuum cleaner hose. The comb, patented in 1952 by Joshua Garner of Englewood, New Jersey, has hollow teeth with ducts through which the dandruff is carried away from the scalp for collection in a closed compartment.

Darts invented by two Michigan men are designed to implant hairs in a bald human pate or in a doll's head. Felix C. Mielzynski of Detroit and Ted Zbikowski of Hamtramck, in their 1961 patent, explain that each dart has a bulbous anchor portion to which one or more natural or artificial hairs are cemented. The recommended tool for implantation is a pneumatic gun with a split needle that is driven in closed and opens to form a cavity in which the dart is embedded. Darts with single hairs are preferred for the part and hairline. (Figure 12.7)

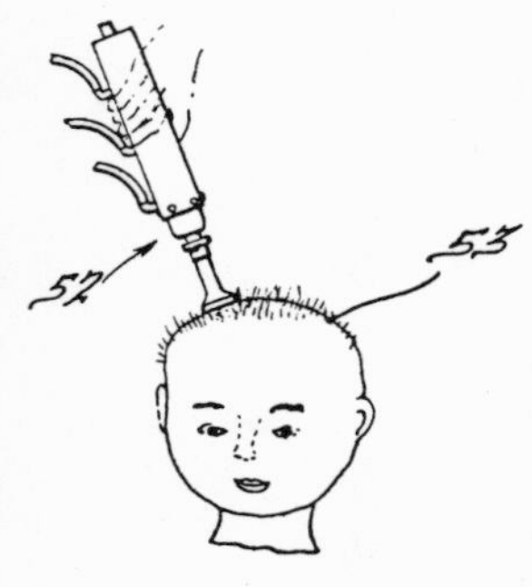

Fig. 12.7:
Doll getting hair implant.
(Patent 3,003,155)

An automated haircut is offered by a French engineer, Jean Gronier of Versailles. The machine he patented in 1966 works from a program recorded for each customer by means of perforations in a strip of material. When the strip is run around a cylinder, the perforations throw switches to control the barber tools. Every time the patron comes in he can get the same trim.

As pictured in the patent, the subject is seated in an armchair with clippers and combs held overhead. He sits still and keeps "feelers" in contact with his scalp.

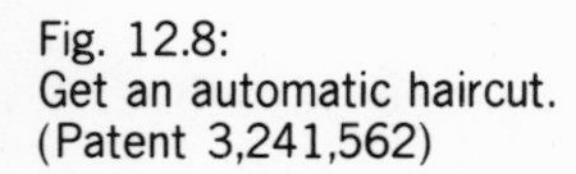

Fig. 12.8:
Get an automatic haircut.
(Patent 3,241,562)

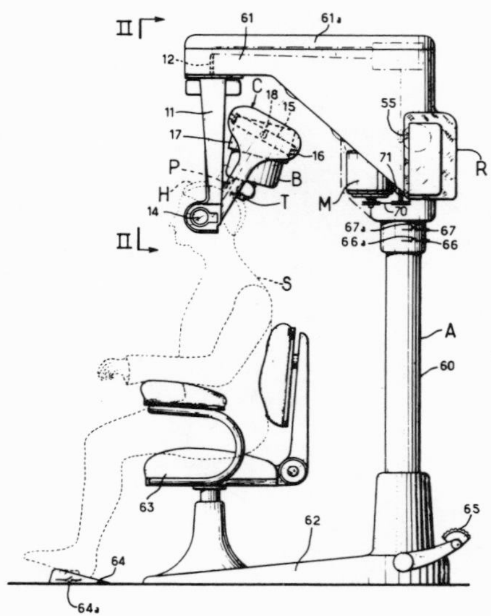

The patent explains that a trimming comb makes a forward raking movement through the hair but in the case of hair that is "cut short in a stubble and brushed straight up from the forehead" the comb is of no use and is held out of the way. (Figure 12.8)

Snoring has long been a challenge to inventors. Early attempts to curb it were divided between the promotion and the prevention of air passage through the mouth. In 1897 Svante Anderson of Chicago obtained a patent for a mouthpiece to be kept in place by an elastic band passing around the head. "Snoring is caused by breathing with the mouth open," he says, "thus allowing a large and unbroken volume of air to enter the glotis *[sic]* ." With his invention, he asserts, snoring is prevented by causing the volume of air to be broken as it passes through slots in the central portion of the mouthpiece. The lips come close together near the corners of the mouth but the central passages remain open and because of their arrangement cannot become clogged by the accumulation of saliva. (Figure 12.9)

In 1930 Richard Garvey of Los Angeles perfected a mouth-closer designed to prevent both mouth breathing and snoring. Made of "any suitable material," it is a solid rectangle to be applied over the mouth, with a rib on the underside to fit between the closed lips and with a pair of straps looped around the ears to hold it in place. Another Californian, Donald H. Waite of San Francisco, disclosed in 1939 an antisnoring device to be held between the teeth and lips. Its perforations "diffuse or break up the air stream, cut down the

Fig. 12.9: Mouthpiece prevents snoring. (Patent 587,358)

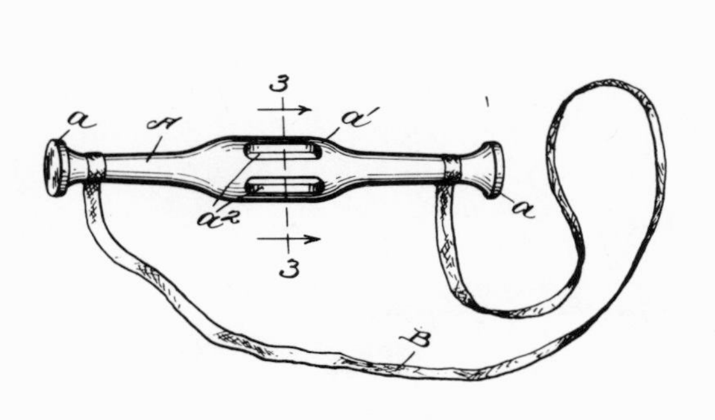
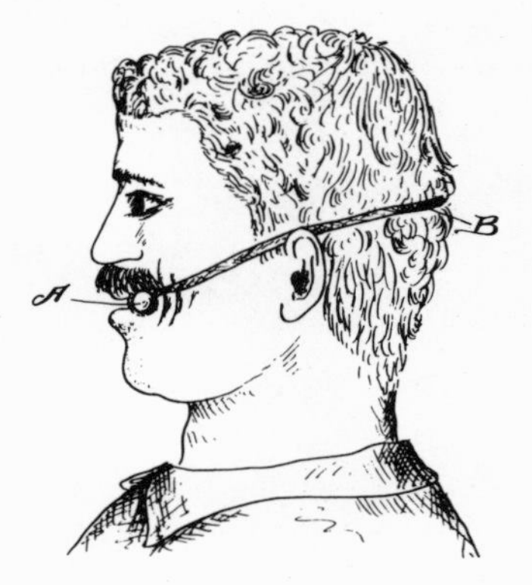

volume of air passing through as well as reduce its velocity, and so aid in the elimination of vibration of the uvula or soft palate." Larger slots permit the normal circulation of saliva. The device is said to encourage the flow of air through the nasal passages, "thereby encouraging good habits as well as correcting and eliminating annoying ones."

In 1953 Elsa L. Leppich of Seattle gave the user of her "dam" a choice of breathing through the nose or not at all. It is a thin sheet of plastic shaped to fit between the teeth and lips with no apertures for air or moisture. According to the patent, the devices also "have been found to be an aid to hearing by correcting some causes of deafness."

A mouth-opening alarm invented by George J. Wilson of Groton, Connecticut "causes the wearer thereof to become conscious of the fact that he has lost control of his mandible," says the 1961 patent. The mechanism is held under the chin by straps that buckle over the top of the head. When the wearer's jaw drops during sleep, that sets off a battery-powered vibrator and an audible alarm. It is predicted that the person will eventually become conditioned to close his mouth while sleeping. (Figure 12.10)

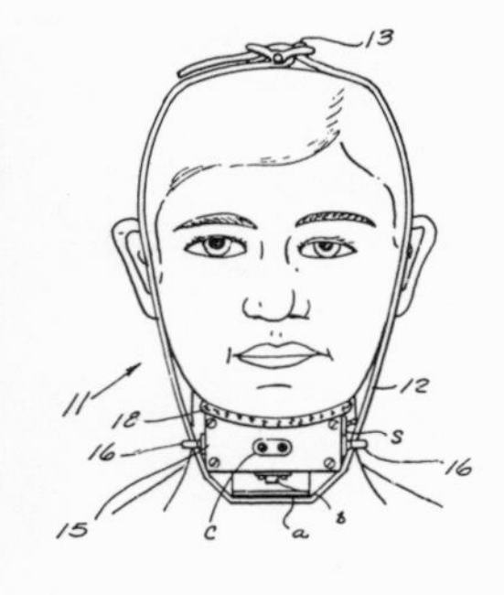

Fig. 12.10:
Open mouth sounds alarm.
(Patent 2,999,232)

Two years later the same inventor patented a snore alarm that rouses the sleeper without (we are told) disturbing anyone else in the room. A microphone is mounted on the head of the bed above the occupant's nose. When it detects a snore, a hinged board under the pillow starts shaking and jarring the occupant. "When the sleeper is thus awakened, he becomes aware of the fact that he is snoring." (Figure 12.11)

Fig. 12.11:
Snore shakes the bed.
(Patent 3,089,130)

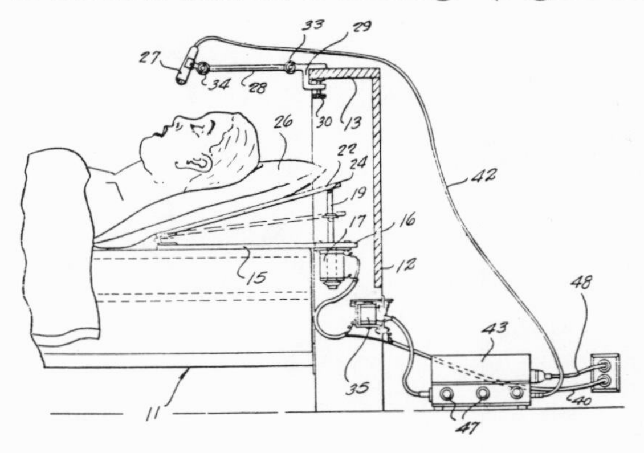

Some big companies are intimately concerned with affairs of the mouth. One is the Wm. Wrigley Jr. Company of Chicago, which in 1966 protected more than 40 formulas devised by Arthur J. Comollo for additives that keep chewing gum from sticking to false teeth. The key ingredient is tannic acid. Some years earlier the company had tackled the problem from another angle—by developing plastic denture material to which gum would not stick—but the denture manufacturers did not take it up.

When even your best friends won't tell you, how do you test your breath? Several methods are recorded in the patent files. George Starr White of Los Angeles remarked in 1925 that "a subject who has been eating onions will not notice the odor of the breath of another who has also eaten onions. A subject under normal conditions does not smell his own breath. I have found upon experiment that if a subject collects his exhalations in a container and then suddenly expels them into his own face or nostrils, he will be able to smell his breath."

Mr. White's halitosis detector is a bellows made of material such as gelatinized paper, with metal ends, one carrying a mouthpiece and the other a ring to serve as a handle. The user puts the mouthpiece between his lips, and repeatedly inhales through the nose and exhales through the mouth until the bellows has expanded. If inhalations were through the mouth alone, malodors from the nasal passages would be missed. With the mouthpiece pointed toward his nostrils, the subject collapses the bellows suddenly, expelling the gases and vapors. That should tell him what his friends won't. The inventor adds that the subject may perfume his breath and test results with the detector.

A breath tester patented in 1957 by Otto M. Dyer, Jr. of Detroit is designed to give an immediate report upon exhalation. The instrument somewhat resembles a stubby smoking pipe. The owner grips the stem between his or her teeth and the bowl fits closely under the nostrils. "The user exhales a short, soft breath, preferably from the diaphragm," says the inventor, and then inhales through the nose. The olfactory nerves should tell whether anything is amiss. (Figure 12.12)

Fig. 12.12:
Self-tester for halitosis.
(Patent 2,780,220)

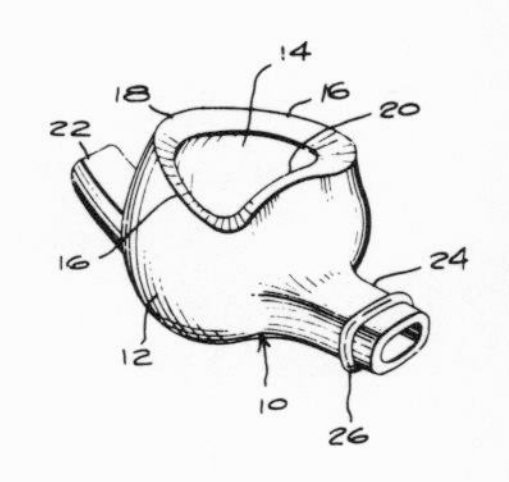

For a pleasant taste in the mouth and a sweet breath, Samuel Henry Sipos of Garden City, Michigan recommends what he calls odoriferous dental apparatus. As described in a 1971 patent, it is a hollow false tooth filled with absorbent material saturated with an aromatic liquid. The patent illustrates it as a molar that forms part of an upper denture. Such pleasant-tasting oils as clove and wintergreen are suggested as producing a sweet-smelling vapor. (Figure 12.13)

Clothing is a matter of deep personal concern to men, and straps often play a key role. Samuel L. Clemens of Hartford, Connecticut, better known as Mark Twain, got an 1871 patent on an adjustable and detachable elastic back-strap for vests, pantaloons and other garments. The straps were to be fastened to the garments with buttons, and their loose ends could be joined with buckles or hooks. (Figure 12.14)

Mark Twain may well have been familiar with the combination necktie and watch-guard disclosed three

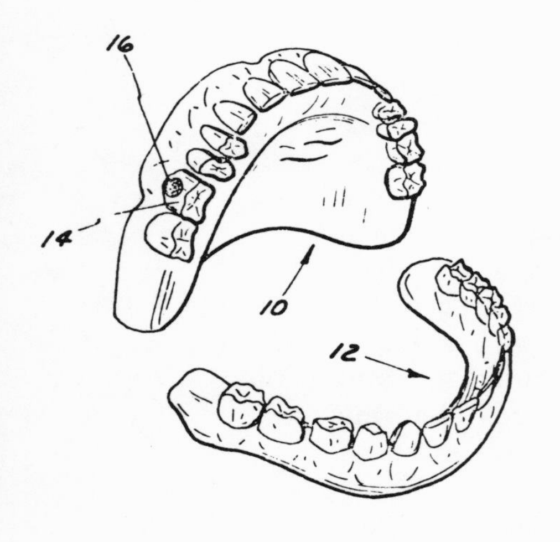

Fig. 12.13:
Sweet odor from a false tooth.
(Patent 3,600,807)

Fig. 12.14:
Mark Twain's pantaloon strap.
(Patent 121,992)

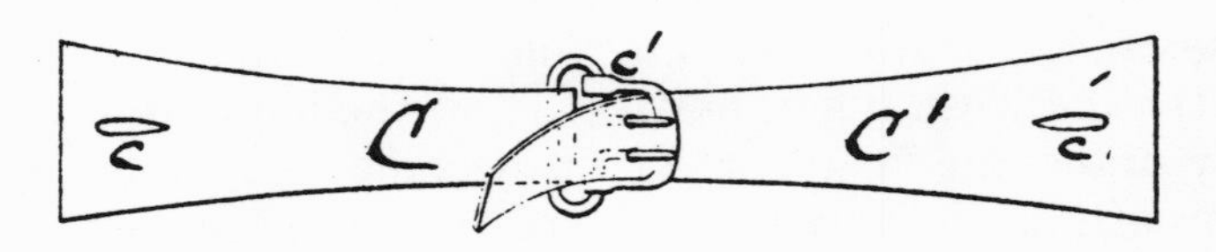

years earlier by Thomas J. Flagg of New York. As pictured, it is a band whose widest part goes around the back of the neck and whose tips are attached to the watch in the wearer's vest pocket. A sliding knot or bow that surrounds the tie is held in place under the chin by attachment to the top shirt button. (Figure 12.15)

A garment for protection against rain, snow and sun is entitled simply Canopy. The hat part, as illustrated in the 1932 patent granted to Asa B. Crosthwait of San Antonio, Texas appears to be about three feet in diameter and is supported from the wearer's waist and shoulders, leaving the head free of weight. The hat may be worn by itself for shade and fresh air, or the waterproof garment may be suspended from it, with a window in front of the face, sleeves for the arms and a skirt providing plenty of play for the legs. The patent does not say, but evidently the Canopy is not intended for wear in a subway crowd. (Figure 12.16)

Fig. 12.15: Necktie guards your watch. (Patent 79,063)

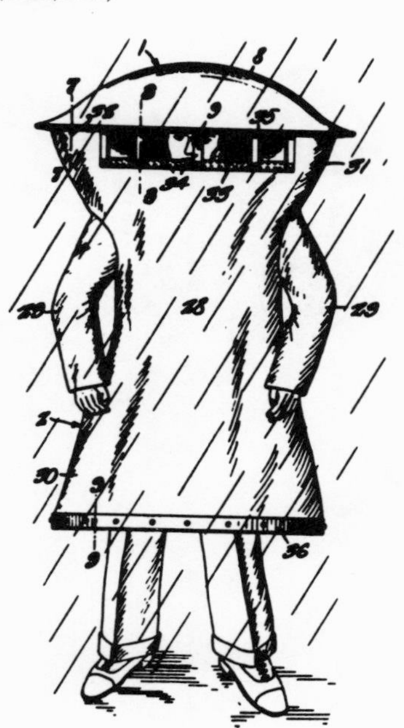

Fig. 12.16: Well-protected from the weather. (Patent 1,888,909)

In a method of fitting clothes by photography, described by Frederick A. Purdy of Scarsdale, New York in a 1953 patent, pictures are taken from various angles of a customer who might want a suit. The front view shows his height, whether his shoulders are square or drooping and the relative length of his arms and legs. A side view reveals where he is hollow or bulging. Both photographs are matched by an operator against patterns displayed on the surface of a large, illuminated drum. The proper suit can then be ordered from stock or from a clothing factory. (Figure 12.17)

Let's say you have plenty of clothes in the house, but how do you decide what to put on for the kind of day it is? Dinko Vrsalko of Split, Croatia, Yugoslavia came up in 1967 with a weather condition indicating device. He remarks that too often people dress by the thermometer—when they should also consider humidity, wind, fog, rain and snow. An outdoor detecting instrument simulates a clothed body with respect to normal body heat and temperature and reflects human reaction to the prevailing weather. Wires run indoors to a dial that is calibrated to indicate the quantity and kind of clothing that a person should don before venturing out.

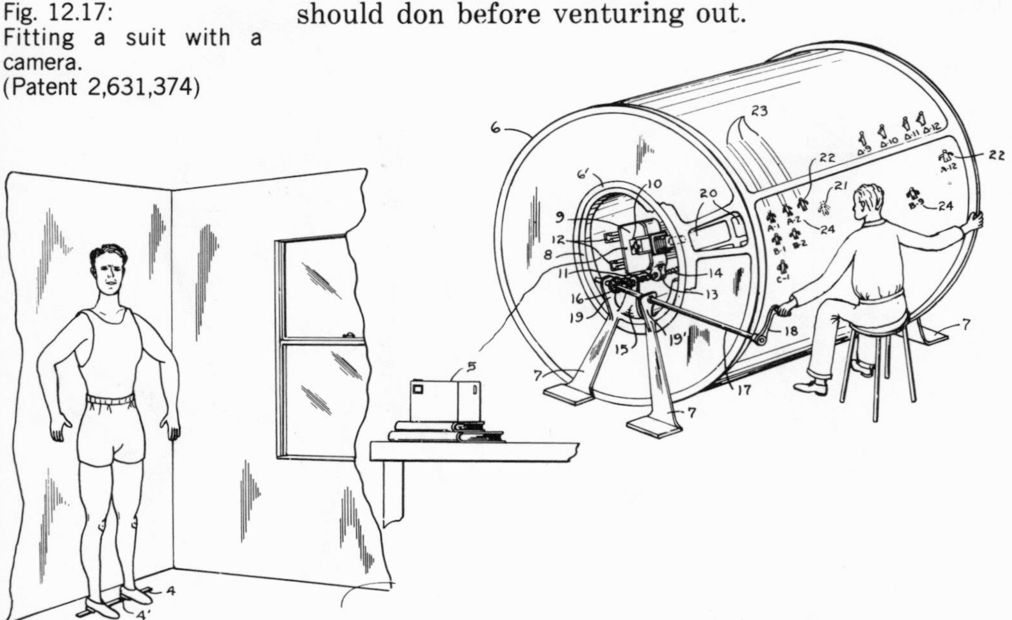

Fig. 12.17:
Fitting a suit with a camera.
(Patent 2,631,374)

As a conversation piece, two Wisconsin inventors disclosed in 1972 a miniature pair of panties for a man to insert in his breast pocket like a handkerchief. Emil V. Beno of Elm Grove and Michael A. Novak of Milwaukee say the lace borders and parts of the legs can be left projecting from the pocket. During the normal course of conversation the wearer will casually pull it out as if to blow his nose, and surprise and chatter will ensue. (Figure 12.18)

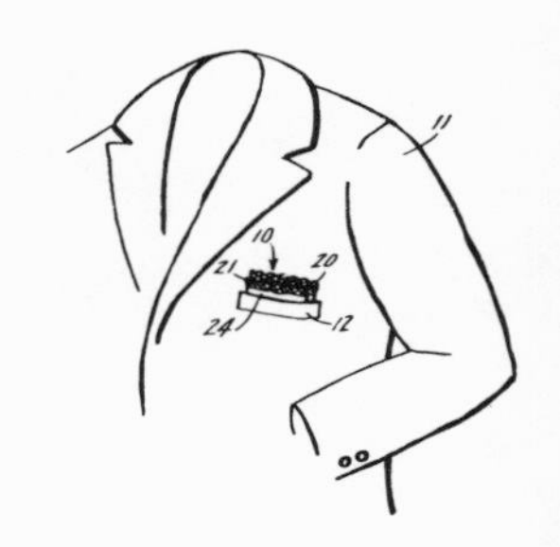

Fig. 12.18:
Panties instead of a hanky.
(Patent 3,704,470)

Man's crowning bit of apparel, the hat, has been improved upon in several ways. Notable is the luminous headgear revealed in 1883 by Robert F. S. Heath of Camden, New Jersey. The hat, cap or bonnet is dipped in or coated with a self-luminous material (not otherwise defined). Mr. Heath says it makes the article easy to find in dark closets and beautiful when worn at night. A luminous hat also helps to spot a wearer engaged in a hazardous occupation, such as a miner or mariner. (Figure 12.19)

A ventilator patented in 1890 by Albert Lee Eliel of La Salle, Illinois has an exhaust fan to be fitted through an opening in the top of a hat and operated by a hand-wound spring. Mr. Eliel explains that the fan draws air in through openings in the side of the crown and induces an upward and outward current, "which has the effect of maintaining a supply of fresh air within the hat and keeping the wearer's head cool." (Figure 12.20)

Fig. 12.19:
A luminous lid.
(Patent 273,074)

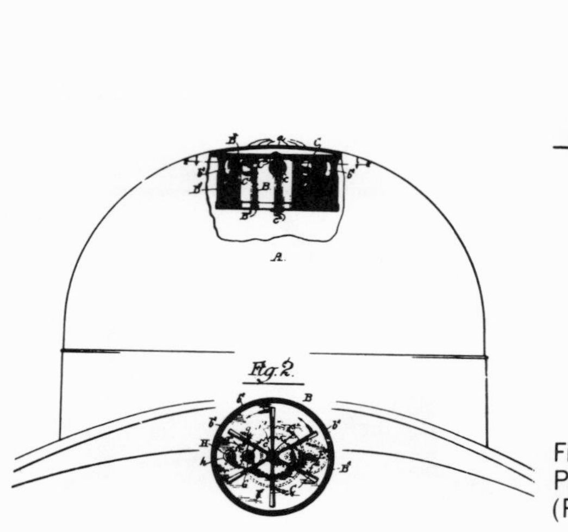

Fig. 12.20:
Pump the air out.
(Patent 432,728)

Harold W. Dahly of Chicago has called on the sun to operate a fan inside the top of a hat. His 1967 patent describes a solar cell that generates current to run the motor. To regulate the speed of the fan or shut it off entirely, a cover can be swung over the cell. Air is admitted through holes in the side of the crown and is circulated for the comfort of the wearer. "It is well known," says Mr. Dahly, "that cooling the top of the head will have a cooling effect on the entire person." (Figure 12.21)

The hat patented in 1892 by William G. A. Bonwill of Philadelphia serves also as an ear trumpet. A wearer who is hard of hearing removes the hat and points its open end at a speaker or other source of sound that he doesn't want to miss. In the open space is a cone, formed preferably of aluminum, with its small end terminating in a hole through the top of the hat. The owner can place this tube end against his ear. In a modified form there is a rubber extension of the tube permitting the user to hold the hat in his lap and listen to the end of the tube. (Figure 12.22)

Fig. 12.21:
Hat cooled by the sun.
(Patent 3,353,191)

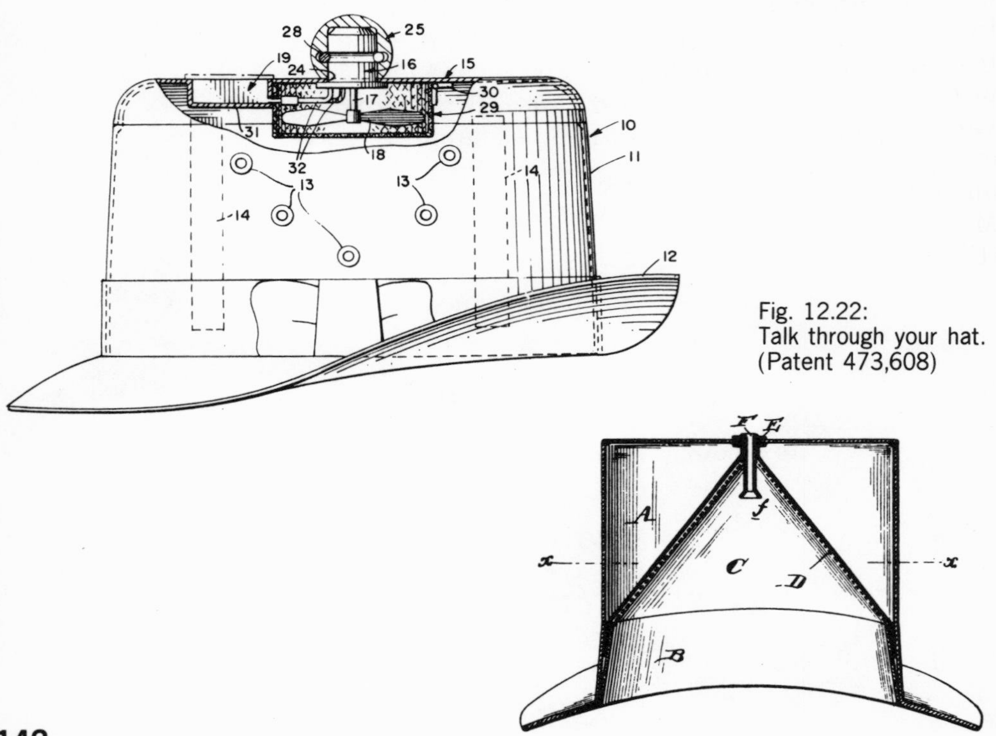

Fig. 12.22:
Talk through your hat.
(Patent 473,608)

Another convenience offered men in the 1890's (1895, in fact) is a napkin holder to be hung from the shirt-collar—letting the linen spread out below—to cover the shirt and coat front. Benjamin Franklin Pascoe of Globe, Arizona Territory made his holder of two attached pieces of metal. At one end they form a pair of tweezers to grip the edge of an open napkin, while at the other end there is a hook to fit in the collar. The tweezers are strong enough so that they can hold a folded napkin together and take the place of the usual napkin ring. (Figure 12.23)

One recent aid to personal care is an illuminated mirror to let a man look into his own ears. Earl M. Christopherson of Seattle patented in 1960 his combined reflector and flashlight. He remarked that it had been difficult, if not impossible, for one to check the internal cleanliness without the assistance of some other person. In the same year Fred A. Williams of Akron, Ohio protected his meditation device. In its simplest form this is a band of metal or plastic, inside which is a hollow strip of foam rubber fitting against the face and forehead to shut off all light, "thereby facilitating contemplative meditation." In modified forms, pictures or symbols can be made visible in phosphorescent paint. A third invention offering advantages in that area of the body is apparatus that enables the user to massage his scalp while standing on his head. Samuel Rubin of Pleasant Ridge, Michigan got a 1965 patent on his exercise device. (Figure 12.24)

Fig. 12.23:
How to hang a napkin.
(Patent 541,384)

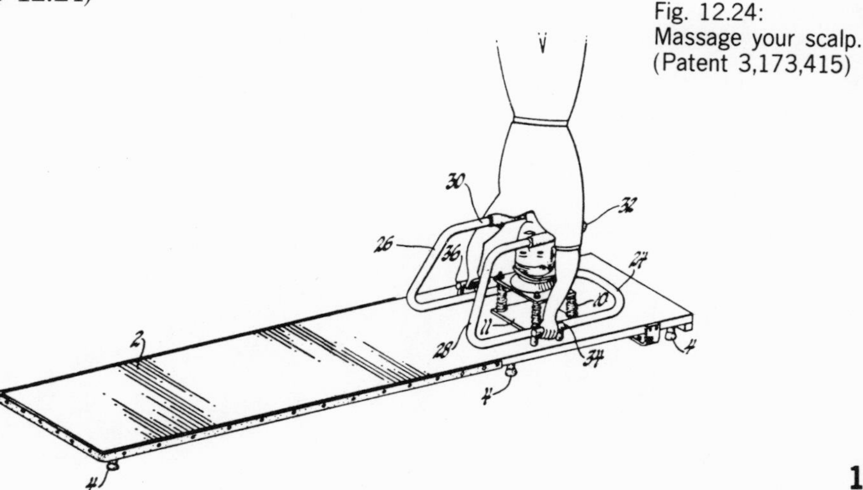

Fig. 12.24:
Massage your scalp.
(Patent 3,173,415)

It may be appropriate to end this discussion of personal inventions with one that is attached to the toes. Russell E. Greathouse of Cuyahoga Falls, Ohio is concerned that in sunbathing, the feet naturally spread apart so that the inner surfaces of the legs get worse burns than the outsides. He was granted a patent in 1973 for a device shaped like a figure 8, which goes around both big toes and holds them together. The result is a relatively uniform burn over both legs. For decoration, and perhaps to give the sunbather something to wiggle, Mr. Greathouse provides a hole in the toe-holder through which the stem of a flower—natural or artificial—may be inserted. (Figure 12.25)

Fig. 12.25:
Hold your toes together.
(Patent 3,712,271)

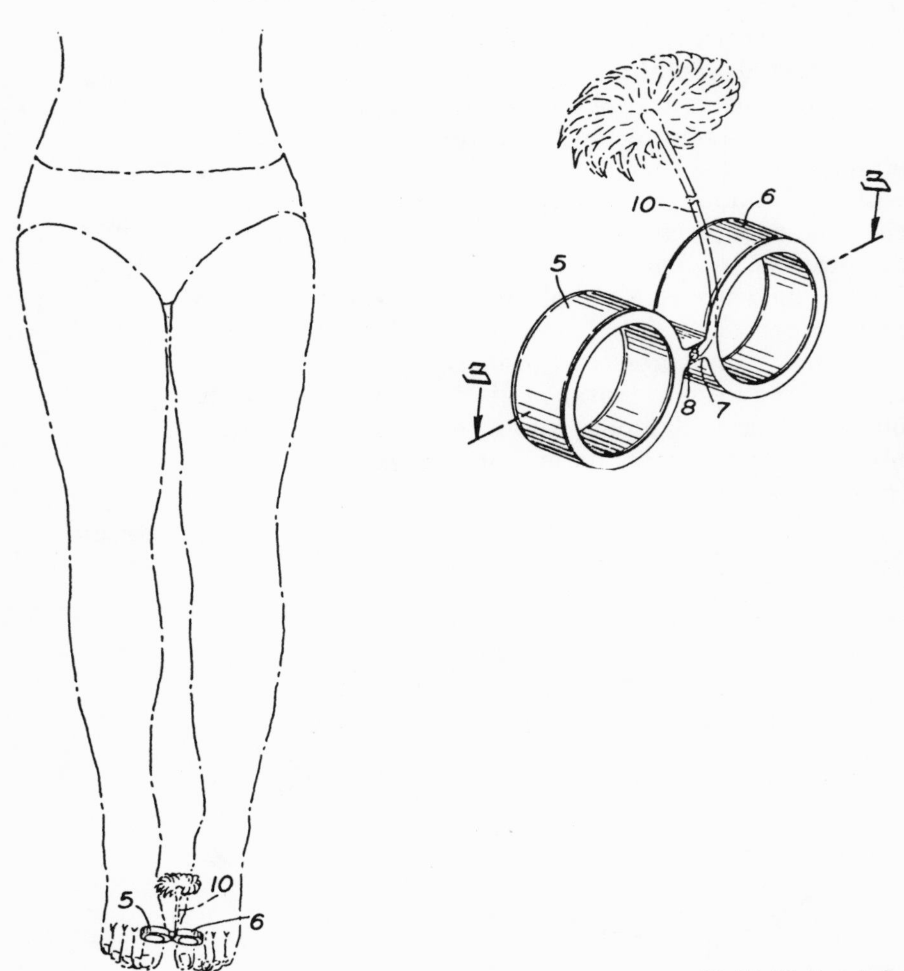

73 74 75 5 4 3 2 1